Is It Time to Abandon Darwin?

New Advances in Scientific Theory

Michael Barber

Is It Time to Abandon Darwin? — Part I

For worldwide distribution

ISBN-13: 978-1-291-94930-8

Copyright © Michael Barber, 2014

The right of Michael Barber to be identified
as the author of this work has been asserted
in accordance with sections 77 and 78 of the
Copyright, Designs, and Patents Act 1988.

Dedicated to
Souvenda, the greatest
physicist of them all;
a good mentor
and a true friend.

Table of Contents
Part I

1	INTRODUCTION	13
	Outline	13
	Why You Should Be Aware of Dynamic Evolution	13
	The Scope of Dynamic Evolution	19
	Dynamic vs. Passive	19
	The Approach and Methodology of this Work	20
	Ultimate Questions on Fundamental Subjects	22
2	CONSEQUENTIAL ENTITIES	27
	Outline	27
	Dynamic Evolution and Quantum Mechanics	28
	The *Big Bang:* Consequential Forces & Objects	29
	The Synergism of Consequential Entities	30
	Water — Simple Example, Complex Reality	31
	Water — a Few Technical Details	34
	A Matter of Consequence	38
3	THE PLEXUS	41
	Outline	41
	Categories of Entities	42
	Examples of Categories	43
	Entities and their Place in the Plexus	45
	Examples of Consequential Entities	48
	Latticework — the Eye-Candy of the Plexus	53
	Compartmentalizing the Consequential Entities	55
	The Plexus — Exquisite Fractal Reticulation	65
4	AN OUTLINE OF DYNAMIC EVOLUTION	71

	Outline	71
	Dynamic vs Darwinian Evolution	74
	Multi-functional Entities of the Plexus	79
	Multi-functional DNA	80
	Multi-functional Endoplasmic Reticulum	80
	Multi-functional Glial Cells	81
	Multi-functional Astrocyte Cells	83
	The Multi-functional Liver	85
	Multi-purpose Neurovascular Development	86
	Peerless Attributes of Plexus Entities	89
5	COHERENCE THROUGH CONNECTIVES	93
	Outline	93
	Configurative Connections — Fractal Patterns	93
	Configurative Patterns of the Plexus	99
	Configurative Connections — Functional Patterns	110
	Tubes — Not Strings!	110
	Muscle Fibers — Tubes Within Tubes	115
6	MACHINE CODES OF THE PLEXUS	125
	Outline	125
	The Languages of The Plexus	126
	Quantum Language Code	129
	Language Code of the Elements	132
	Language Code of DNA	133
7	THE MASTERMIND EFFECT	139
8	GLOSSARY	161
9	BIBLIOGRAPHY	171
10	INDEX	185

Part II

THE PLEXUS — EXQUISITE FRACTAL RETICULATION

PLEXUS ENTITIES — DISSECTING PROPERTIES AND ATTRIBUTES

WATER, WATER EVERYWHERE — BUT NOT A MOLECULE CAN WE CONTROL

NO BODY PARTS, DR. FRANKENSTEIN

BONE UPON BONE — REACTIONARY PLANNING

CONGRUITY VS. INCONGRUITY OF PLEXUS LEVEL ENTITIES

SYNERGISTIC PATHWAYS & FILIAL CHARACTERISTICS

Part III

THE QUANTUM WORLD — BIRTHPLACE OF DYNAMIC PRINCIPLES

THE HUMAN BRAIN — AN INFINITY OF QUANTUM PROBABILITIES

MACHINES OF THE PLEXUS

PARADOXES WITHIN PARADOXES — A NEW LEVEL OF CONNECTIVITY

THE ECCENTRICITY EQUATION

THE ARTICULATE ARCHITECT

DYNAMIC PRINCIPLES — HOW DO THEY WORK?

"Two things are infinite: the universe and human stupidity;

and I'm not sure about the universe."

— Albert Einstein

INTRODUCTION

1 INTRODUCTION

Outline

1.1 Dynamic Evolution takes you from the world of the sub-atomic, via quantum mechanics and through the progressive stages of size and time to galactic superclusters, demonstrating at each stage a specific, confined set of laws and principles that provide consistent mechanisms for the gradual development of all the objects and forces in the universe. In contrast to the precepts of Darwinian Evolution, its principles will be seen as a "better fit" in terms of providing the answers to the origin of, not just "the species," but of all that we see and discern around us. (Note that this is a work of science, referencing the most up to date findings of contemporary scientists, and does not cover philosophical, religious, or speculative studies.)

Why You Should Be Aware of Dynamic Evolution

1.2 Around the middle of the nineteenth century Charles Darwin released his work *On the Origin of Species*. In a relatively short time his theory was accepted by the scientific community in general. Scientists on the whole adopted the theory that the principle of *survival of the fittest* explained the arrival of the diverse forms of life on earth. However, was the adoption of all the implications of this theory in any way hasty?

1.3 By way of illustration: As any parent would readily agree, children gain a little knowledge and then they seem to  think they know everything. No doubt you have come across many a child who thinks he/she knows all about the facts of life? They have an "of course I know all about it," attitude. But as any married adult is aware, not even a four year university course could hope to cover "all about it." The complexity of the relationship between a man and his wife and the physical, mental, and spiritual depths that they each transcend in the most intimate of physical human experiences, has been the subject of poetry, literature, art, theater, fictional stories, movies, not to mention the abundance of scientific material by psychologists, sociologists, neurologists, anthropologists, etc. And yet the little boy looks up at you when you start on your long, well-planned soliloquy on the facts of life, and rolls his eyes in an "I already know all about it" gesture.

1.4 Of course, adults have many years of experience that children lack. Additionally, mankind has the advantage of accumulated scientific knowledge over many centuries of time. But in a universe as vast, inveterate, and comprehensive as the one we find ourselves in, what does even this accumulated knowledge amount to compared to the sum total of unanswered questions, from the infinitesimally small objects within the atom to the unimaginably big galactic superclusters? Although, for many, mankind stands upon a great "threshold of ultimate understanding," nevertheless, this great understanding is only a "thread" that unites popular theories; it is not (nor does it claim to be) the sum total of universal scientific knowledge!

1.5   Sir Isaac Newton was known for exhibiting a certain "arrogance" due, at least in part, to his obvious intellect. However, he is quoted as saying:

> "I seem to have been only like a boy playing on the seashore, and diverting myself now and then, finding a smoother pebble or a prettier shell, whilst the great ocean lay undiscovered before me... If I have been able to see further it was because I stood on the shoulders of giants."

1.6 Such is man's insatiable quest for enlightenment and truth, even scientists, who pride themselves on their cerebral, logical approach to the acquisition of knowledge, at times take shortcuts to conclusions they at once favor but later regret. Darwinian Evolution appears to be an example of just such a shortcut. Does it provide *all* the answers to the origin of life on earth, to the methods by which the complexities of living organisms came about? Is it sufficiently free from flaws to provide a fully convincing argument? Is it free from the criticism of influential opponents? Darwin named his work "Origin of *Species*" not "Origin of *Life*," nor "Origin of the *Universe*." He noticed how certain characteristics of living organisms change, adapt, over time (as will be seen later in this work, **the passage of great amounts of time turns out to be counter-productive for the Darwinian model** rather than otherwise). But he did not endeavor to tackle the progressive development, over time, of *all* aspects of life, and his theories did not attempt any explanation of the origin of other entities within the universe.

1.7	With regard to the viability of the theory in explaining the origin and development of all the complex systems that work together to complete a living organism, there are many deficiencies. Even its most ardent supporters would freely point out its shortcomings and examples of each specific hiatus. Some may not be very keen to surrender all of its failings, though. The theoretical "model" presented by Darwin has many attractive facets, and the potential to provide many explanations that scientists seek. Therefore, despite its limitations, it clings on relentlessly, forever threatening to find the required support with elusive connecting facts.

1.8	Until something superior, more attractive, more in line with experimentally proven scientific data, gains ascendancy in the minds of the leading lights of the world's intelligentsia, Darwin, it seems, is all we have.

Or is it?...

1.9	**Dynamic Evolution** offers a comprehensive challenge to the Darwinian premise that the diversity of life is **explained solely by the natural selection and survival of the fittest**. It will become apparent from a complete study of this work that the principle of the selection of *strong* functions/properties of life and living organisms, whilst providing a partial fit and a compelling argument, falls under the shadow of the complete picture provided by Dynamic Evolution. Just as a large multi-piece mosaic can appear quite beautiful and coherent, even when some of the pieces do not quite make the right fit (they "look almost right"; the emergent patterns can still be impressive and instructive), **the picture they present seems to lack too many pieces to provide a complete picture**. If the saying "a picture paints a thousand words" is true, then this picture would have too many incomplete or missing sentences. A comparison with the more accurate paradigm, with the pieces in elegant continuity exhibiting many contiguous patterns, presents a more attractive arrangement and a more influential argument.

1.10　The picture presented by Darwin's theory therefore exhibits many non-contiguous patterns, not to say a good many missing pieces. **Recent scientific advances and findings** in fields such as human biology, quantum mechanics, anthropology, and archaeology, have opened up new lines of theory, closed some doors on long-standing theories, and left many other questions unanswered.

1.11　This work endeavors to demonstrate that Dynamic Evolution, by virtue of the coercive argumentation used and the fecundity and uniformity of the inter-connected principles covered, **provides a greater number of "correctly fitting pieces" of the proverbial mosaic**. Indeed, the challenge that this work presents is for the reader to find pieces that cannot be made to fit, or to point out patterns that are inconsistent and can be demonstrated to have a better fit. If a *small* number of "patterns" are found to require a *better fit*, then modification of the theory is required. If a *large* number of patterns are found to require a *better fit*, **then it may be time to abandon the theory**. That said, Dynamic Evolution is concerned less with theory than with available scientific data. Its strengths lie in assembling these facts into a logical arrangement that assists in highlighting advanced principles not available from a study of the Darwinian model.

1.12　Admittedly, some of the concepts described in this volume require a good measure of study and some abstract thinking. However, the extent to which the reader puts forth the effort to acquire an understanding of these concepts, to that extent the benefits will be reaped. It is undesirable to insist on the relentless pursuit of a theory which has had the might of the world's intellectual energy invested in it, and yet is not able to "tick all the boxes" that the full erudition of Dynamic Evolution provides.

1.13　For many decades scientists have searched for the "grand unified theory of everything" (ToE), which would provide unifying principles and formulas to guide researchers, using existing or *modified* prevailing theories, toward a more accurate and complete understanding of how the universe and, in particular, life arrived in its current state. This is often described as a **consolidation of Einstein's theory of relativity with the principles of quantum mechanics**. The principles and theories set out in this work have the noble objective of laying down the groundwork for the "Theory of Everything."

The Scope of Dynamic Evolution

1.14　As stated above, there are a number of differences with the principles and theories presented here compared with the standard model of Darwinian Evolution. One key difference is the *scope* of Dynamic Evolution, which covers the development of the universe from the Big Bang onwards, and all physical and consequential objects, forces, conditions, and laws that result from that development. Hence, the key term *Consequential Entities* is used many times throughout this work and is of necessity used in this generalized way.

Dynamic vs. Passive

1.15　The principles explained in this work have been granted the descriptive name "dynamic" by virtue of the evidence presented in the progressive modules, and the nature of the evidence. This is **contrasted with Darwinian Evolution which is seen to be a "passive" explanation** to the evidentiary argumentation. For example, natural selection, by definition, requires that an organism or function is already present in order to be selected for survival. The Darwinian principles tend toward the strengthening of a function that is in competition with another function for the perpetuation of the organism. But there is **less emphasis on the mechanisms that explain the development** of the competing functions prior to selection; *survival* has been enhanced at the expense of *arrival*. This underscores the *quiescence,* the *passivity,* and necessarily the inadequacies of the Darwinian model.

1.16 **Dynamic Evolution, on the other hand, deals with the processes that underlie the development** of the function or organism. These processes are dealt with in detail and contrasted with parallel or similar developments that form a repeating pattern throughout the Dynamic model. Hence the term "passive" or "quiescent" is contrasted diametrically with "dynamic" or "active" in relation to the very processes that Darwin felt served to underpin his theory.

The Approach and Methodology of this Work

1.17 The concepts covered by this work reference the scientific fields of **quantum mechanics**, **physics**, **chemistry**, **mathematics**, **astronomy**, **biology**, **anatomy**, **physiology**, **zoology**, and **cellular microbiology**. Relevant studies in these fields are cross-referenced and contextualized for each of the arguments that are presented in the study of **Dynamic Evolution**.

1.18 The study itself takes a progressive approach; an overview of the fundamentals is covered first, and references are made to more advanced topics that are dealt with in later sections which **provide a modular approach** to the advanced details of the study, enabling a full grasp of the major features.

1.19   The advanced ideas proffered by this work demonstrate that Dynamic Evolution is the most convincing and attractive alternative to the theory of Darwinian Evolution to date.

1.20   This work takes a top-down approach to the material in the multiple layers of the study of Dynamic Evolution, but it employs a **multiscalar** (and not exclusively macrocosmic) methodology; i.e. its main argumentation begins conceptually from a broad-based view and then scales down into ever deeper, increasingly magnified levels of **compartmentalized analyses**. For example, take the **Plexus** section that covers the subject of Molecules (Level 2): as the analysis progresses, smaller homogeneous studies are introduced that support the principles, explaining how the attributes of the many *Entities* of a certain molecule elucidate the arguments and then reach out to related fields of science before proceeding to an analysis of the components of those enumerated *Entities*, and the cross-referenced principles that expand on and relate to the ideas under discussion—another scale in the cubic-model approach.

1.21   It will become apparent that the scientific data that support the principles of Dynamic Evolution are far more plentiful in the related fields of science than are the supporting pillars of Darwinian Evolution. **The "mosaic" presents a more coherent picture in the *dynamic* model than in the *passive* model of Darwinian Evolution**.

1.22   There are a number of reasons for the above-mentioned methodology and approach. One in particular is the desire to counter the subjective microcosmic view, promulgated by some intellectuals, of highly-ordered life; for example, the Ping-Pong debates of some camps in the presentation of very detailed arguments that endeavor to prove that a particular microscopic facet of a tiny organism did or did not arise by a particular mechanism. Clearly, it is easy to lose sight of the overall point—the macro view, the parental compartment in which the argument under discussion lives—when one's vantage point is at the other end of a microscope.

1.23 Using its *multiscalar* approach, this work succeeds in securing the logical, legitimate place of a given argument within the framework of the **Dynamic Model** by its proximity and cognation to other verified, closely-related or cross-referenced studies which, by contextual validation, reinforce the consistency of the foundation for the intellectual acceptance of that argument.

1.24 The above explanation is in good company. **Einstein once wrote**:

> "I want to know how God created this world. I am not interested in this or that phenomenon, in the spectrum of this or that element. **I want to know His thoughts, the rest are details**."

1.25 Please note that this work does not attempt to explain the "who" "what" "when" "where" and "how" of Einstein's "God." Nor does this work have any affiliation with the proponents of "Intelligent Design" or "Creationism." The implications of the principles of Dynamic Evolution are left to the reader to extrapolate as he or she wishes.

Ultimate Questions on Fundamental Subjects

1.26 The principles of Dynamic Evolution satisfy an age-long thirst for the truth about "this world" that Einstein spoke about in the above quotation. What is it really made of? What is there at the deepest level of the tiniest particles of matter? Is there an "ultimate" (unbreakable) particle that every other particle is made from? Why do the fundamental forces and principles that govern the behavior of these particles operate in the "peculiar" ways that they do with such a high level of consistency? What were they like during and just after the "Big Bang"? How did electrons come about, and what is the secret of their strange behavior? Why is there a high level of uniformity and consistency about their physical form and behavior that is so unlike any other object in the universe? How did sub-atomic particles come about and what was their nature and behavior before they coalesced into atoms? Why does the strong nuclear force only operate at precisely the distance required to cover all stable atomic nuclei and no more (mass of the gluon)? Why is it that quarks are the only particles of matter that interact with all four fundamental forces? What is the nature, composition, and origin of *time*? Why are galaxies, moon orbits, and solar systems mostly flat or disc-like, but atoms, galaxy clusters, stars, and planets are spheroidal? More interestingly, could a greater understanding of particle physics, of DNA, of the fine-tuned forces latent within every element and present in the universe everywhere around us, indeed of the finer principles of Dynamic Evolution, provide answers to mankind's search for sustainable energy and other vital quests?

1.27 To use an illustration: this "quest" is like an investigator overturning stones at an archaeological dig, not just to get to the bottom of the era the site has been categorized in, but to uncover *all* the stones, artifacts, debris, surrounding material, etc., in a bid to reach the limit of the understanding that is possible for that site. **The claim of this work is that Dynamic Evolution, when followed through to its conclusion, fully satisfies this quest**.

1.28 The questions formulated in §1.26 can be satisfied by a study of Dynamic Evolution. The principles described in this and subsequent volumes in the series, demonstrate an elegance, power, and beauty manifest in nature and in the universe around us that can provide answers to vital, centuries-old questions. Could a greater understanding of Dynamic Evolution provide answers to the vital questions referred to above, indeed answers that lead to the ultimate "Theory of Everything"? This power, these latent forces, need to be harnessed to be of use; they must be understood before they can be harnessed; but **they must be appreciated for their place in the overall scheme of things** in order to be investigated efficiently and understood adequately. And it is this "grand scheme," **the coalescing and ultimate comprehension of synergistic "Dynamic" principles that have the potential to lead ultimately to the required understanding**.

1.29 It is therefore readily apparent that it is of great importance that Dynamic Evolution itself is propounded and understood. As its advanced rules and precepts are dealt with, it will become clear that many hidden treasures can be discovered and extricated using *Dynamic* principles. Moreover, the inadequacy of Darwin's Evolution will become more obvious, and associated **scientific endeavors can at last be turned toward more successful lines**, each possessing the potential to considerably benefit mankind's future.

Consequential Entities

2 CONSEQUENTIAL ENTITIES

Outline

2.1 In order to demonstrate the beauty and elegance of the mechanics—the working principles—of Dynamic Evolution, it is necessary to explain how a multitude of facets, or properties, of forces, conditions, and objects in the universe are subject to certain consistent patterns of performance and form, that vary little or at all from the laws that govern their structure and behavior. These are termed **Consequential Entities**.

Figure 1 - Basic illustration of Entity connectivity within the Plexus

2.2 Scientists, on the whole, now believe that all the matter in the universe was formed as a result of the Big Bang. Therefore, everything that occurred and that came into existence after the Big Bang is a Consequential Entity.

There are fundamental forces that have a direct relationship with the constituents of matter; for example the strong and weak nuclear forces, and gravity. These are also classed as Consequential Entities as they have specific values and parameters that only vary according to given laws and conditions that are themselves subject to the principles of Dynamic Evolution.

2.3 Figure 1 illustrates how, given that objects (the spheres) represent the Entities within the Plexus, there are multiple connections possible. It will become apparent as you make progress through the study of Dynamic Evolution, that the sheer number of connections between Entities is raised to a mind-boggling level of complexity.

2.4 Examples are given below of Entities and Entity types. It is beneficial to study the concepts carefully, as these are vital to a complete understanding of the advanced principles of Dynamic Evolution.

Dynamic Evolution and Quantum Mechanics

2.5 One example that is causing considerable excitement for physicists is the discovery of the **strange yet persistent rules** that govern *the behavior of the movement of sub-atomic particles, atoms, and molecules*—the field of quantum mechanics. This branch of physics has undergone considerable change in the past few decades; in particular, **the laws that govern the behavior of these particles and elements have become many times more interesting** than the "classical" principles scientists used to apply exclusively to them. Furthermore, some of the principles that make the study of quantum physics so fascinating find their parallels and counterparts at the higher levels of **The Plexus** (discussed later in this study) and underscore the solidity of the principles of Dynamic Evolution.

2.6 The Consequential Entities, these "forces and objects," can best be described as all the processes, elemental objects, recombinant objects, conditions, and forces that exist and have existed in the universe, and that are therefore related, **bound together by a consistency and a logical arrangement** that are best explained by Dynamic Evolution. Moreover, the compelling explanation of these Entities **can only be delivered by a complete study** of the principles covered by this work, as the principles explained within are expounded and expanded upon from the sub-atomic level upwards.

2.7 **Quantum principles themselves exist as a recurring theme throughout this study.** Their association with sub-atomic particles, atoms, molecules, proteins, light waves, photosynthesis, brain functionality, star formation, etc., is seen in the context of the consistent principles of Dynamic Evolution.

The *Big Bang:* Consequential Forces & Objects

2.8 As discussed, scientists believe that all the matter in the universe was formed as a result of the Big Bang. Therefore, **everything that occurred and that came into existence after the Big Bang is a** *Consequential Entity.* Some of these are also *Fine-tuned Entities.* This enables a key concept, namely *dependencies,* to be enlarged upon when discussing the principles of Dynamic Evolution, without limiting the study to only those Entities that have some fine-tuning parameters.

2.9 The "fine-tuned entities" bare the following description:

> Fine-tuning relates to the various processes, elements, and conditions that exist in the universe with specific parameters (sizes, speeds, strengths, distances, etc.) that only work effectively (or at all)

if those parameters are within a particular value threshold; for example, plus or minus 1% (+-1%).

The Synergism of Consequential Entities

2.10   The universe is composed of an unknown but considerably high number of these Entities. However, an examination of the chapter **Dynamic Evolution** reveals that these objects do not *all* fit into the category of *Fine-tuned* Entities. Many objects are not actually fine-tuned in any particular or discernible sense; they are capable of subsisting without any exact parameters being applied to them, and of forming "normal" connections with other Entities within the Plexus.

2.11   For example, if a certain type of rock had a different consistency, it is unlikely to affect the existence or continuity of anything else; the term fine-tuning could not effectively be applied here; unless of course a rock suffered a collision with another object for example in the asteroid belt and became bound on a collision course with earth! But, the synergism between the majority of these Entities demonstrates that there is an **unbreakable bond or cohesion manifest between the components of the Plexus**. *Synergism* is described by the *Merriam-Webster* dictionary as: "interaction of discrete agencies (as industrial firms), agents (as drugs), or conditions such that the total effect is greater than the sum of the individual effects." This therefore fits in well with the relationship between the Entities, and the *dependencies* that are formed: similar to a stack of dominos placed edge-on each one in close proximity to the next. If just *one* domino topples, they all topple. Hence, reference is frequently made to the *domino effect*. Later chapters in this work demonstrate the application of this principle to the multiple **Entities** within the **Plexus**.

2.12   Therefore, how many of the objects, conditions, and forces form a *dependency* with other objects, conditions, and forces? If a certain object just happened to "fail to come into existence," would it affect other entities? **If oxygen never existed, how could water exist?** If oxygen did not exist, how could blood come about? The strong nuclear force belongs to the Fine-tuning category, but what if it didn't exist at all? The essential binding in the nucleus of atoms could never occur. Molecules would never form because there would be no atoms more complex than hydrogen. Therefore, no elements of any value would exist, let alone elements essential for star formation, planetary nebulas, and for life. What if the electron never came into existence? How would elemental bonding, radiation, ions, or indeed most chemical reactions, occur?

Water — Simple Example, Complex Reality

2.13   The *dependencies* between Entities within the Plexus can be illustrated using the example of the apparently simple water molecule. This consists of two atoms of hydrogen and one atom of oxygen. The properties of the water molecule, however, are manifold. These will be developed in more advanced studies; at this point, water will serve to illustrate how the concepts of Dynamic Evolution are applied in specific ways to objects and forces from the tiny atom to the largest Galactic Supercluster and are best illustrated by The Plexus within this context.

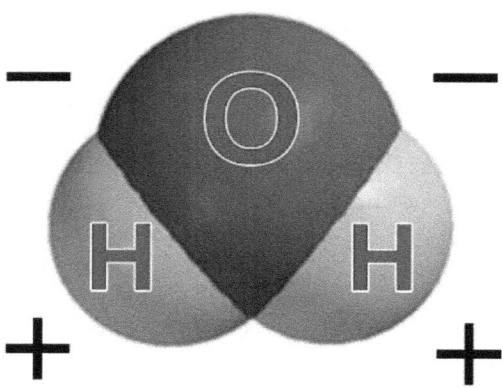

Figure 2 – Artistic arrangement showing a single water molecule

2.14   The above illustration shows two molecules of Hydrogen (H^2) bonding with a single molecule of Oxygen (O) to form the water molecule. Electrical attraction is also involved in at least two basic ways: (1) the molecule's own firm particle binding, and (2) the way the molecule is able to bind to other substances. These three *Entities* (electrical conductivity, hydrogen, oxygen), are a small sample of the many objects and forces that water depends on for its existence and continuity. Therefore, the following illustration is a tiny simplified fraction of The Plexus, demonstrating connectivity between these Entities:

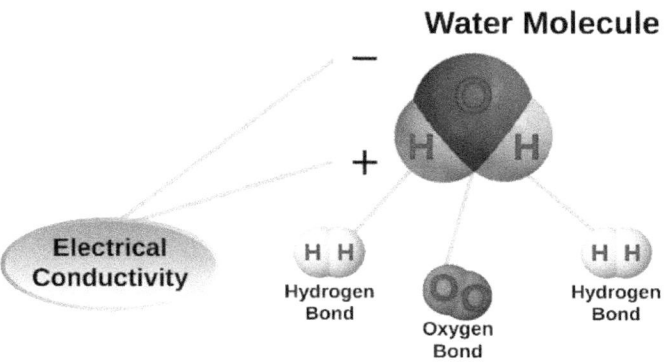

Figure 3 - Some of the dependents of the Water Entity

2.15 In this illustration, a very small portion of the Entities that water depends on are shown. There are many more Entities in the opposite direction, *dependents*.

2.16 As mentioned above, H_2O is sometimes referred to by scientists as a "simple" molecule. However, note this partial list of the properties of water:

- High cohesion
- Adhesion
- Surface tension
- Heat capacity (high heat conductivity)
- High latent heat of evaporation
- Intra-molecular hydrogen bonding
- Finely balanced miscibility (ability to mix)
- Very light as a gas
- Very dense as a liquid
- Hydration (which has many sub-sets)
- Ionization factors
- Density-driven thermal convection
- Expansion on freezing

- Hot water freezes faster than cold water
- Peculiarities of behavior under differing pressures
- Brine rejection (salt distribution)

2.17 Each of these items, plus the additional items in the full list (dealt with in later studies), constitute the *Properties* of the Entity **Water**. Many of these have their own respective connections to other Entities within the Plexus. One simple example is the 'Expansion' property, where **water molecules expand as they approach freezing point**. A major dependent of this property is earth's marine life (a collective Entity), which would not survive the permanent freezing of vast oceans and seas. Another example is photosynthesis, a far from trivial process involving **chemical, electromagnetic, and even quantum processes** to convert sunlight, carbon dioxide, and water into oxygen and carbohydrates.

2.18 Therefore, just as water depends on other Entities, certain Entities also depend on water for their continuity. This is a dramatic example demonstrating a few of the advanced principles of Dynamic Evolution.

Water — a Few Technical Details

2.19 Water is composed of just two atomic elements: hydrogen and oxygen. Two atoms of hydrogen bond with one atom of oxygen to form a water molecule, hence the term H^2O. The nucleus of a single atom of hydrogen normally has a positive charge of one unit and is balanced by one electron making the atom electrically neutral. Two hydrogen atoms bond with one oxygen atom to form a banana-shaped molecule with oxygen forming the fat middle section. As the electron orbit in hydrogen is smaller than that in oxygen, the electrons that orbit both the oxygen nucleus and one of the hydrogen nuclei is more often to be found near the oxygen atom, giving the molecule a negative charge near the "bend" of the banana where the oxygen atom is, and a positive charge centered opposite the "ends", i.e. between the two hydrogen nuclei.

2.20　The electrical charges ensure that water molecules are attracted to each other. They form long, fragile chains that are too heavy to move freely at low temperatures; therefore water is a usually a liquid. Although apparently simple, **water requires an advanced knowledge of physics to understand all of its properties and dependencies fully**. Note the following basic facts:

- If water was not electrically polar, it would be a gas at room temperature and have an extremely low freezing point, making life impossible. Because of the shape of the water molecule, the distribution of the electrical charge is asymmetric. The oxygen nucleus draws electrons away from the hydrogen nuclei, which leaves the region around the hydrogen nuclei with a net positive charge. The water molecule is thus an electrically polar structure.

- The polar nature of water enables it to form a 'skin' over the surface of a body of water, strong enough to support tiny, light objects. This is known as **surface tension**. Water has the highest surface tension of all common liquids. Water is therefore highly cohesive. Water molecules interact strongly with one another through hydrogen bonds. These interactions are apparent in the structure of ice. Networks of hydrogen bonds hold the structure together; similar interactions link molecules in liquid water and account for the cohesion of liquid water, although, in the liquid state, some of the hydrogen bonds are broken. The highly cohesive nature of water dramatically affects the interactions between molecules in aqueous solution.

- Water has a great capacity to hold heat energy, with the highest heat of vaporization of most common substances (thus a high boiling point—enabling it to remain as a liquid on the surface of the warm Earth). When water evaporates, it absorbs considerable amounts of heat.

- Water has a high latent heat of fusion; when ice is formed, considerable amounts of heat energy is released. Water therefore acts as a buffer against temperature changes and **keeps earth's climate from rapidly fluctuating**.

- As considered above, when water freezes, it becomes less dense—hence ice floats (if this were not so, the oceans would be frozen solid).

- Possibly most important for the chemical processes of life: **water is a universal solvent**. It has the ability to dissolve more substances than any other liquid (due, once again, to its polar characteristics and hydrogen bonding). When dissolved in water, salts turn into the ions Sodium chloride. Table salt, NaCl becomes Na+ and Cl–. This allows for many free radicals to be available for the chemistry of life.

- Water is very dense, some 800 times denser than air. This density allows large and small organisms to float along effortlessly for long periods of time (compared to land, where terrestrial life must counter gravity with each step in order to move.)

- **Water absorbs light rays very quickly** (important to photosynthetic life, which is only possible where light penetrates, and light is absorbed as deep as 600 feet beneath the surface of the oceans).

- **Water absorbs light differentially.** The red end of the light spectrum is absorbed in shallow water while the blues and greens penetrate the deepest (important for plants because different plants use different parts of the light spectrum for photosynthesis, and the differential absorption can determine the vertical distribution of marine plants).

2.21 Some of the above principles can serve to illustrate the

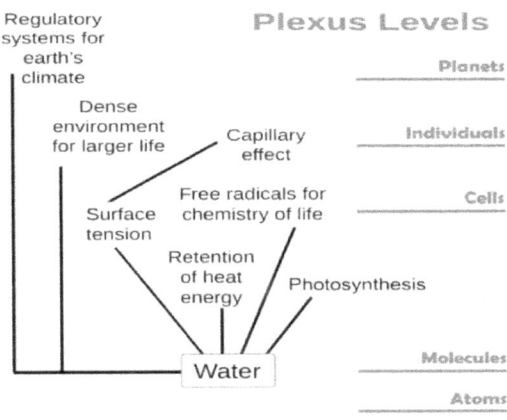

Figure 4 - Greatly simplified sub-section of Entity connections to water

nature of *The Plexus*, and **the distinct ways in which Entities are related to one another at all levels**. We have seen a simple example of the Entities upon which water depends. Now in the opposite direction, note the simplified illustration (above) of the Entities that depend upon water.

2.22 It can be seen that the Entities depicted in this illustration, and their basic connections, are presented in the general order in which the levels of The Plexus are described (see next Chapter). This **concept of Entities which *span* levels** is one of the more fascinating aspects of this study and compels further investigation in the more advanced stages.

2.23　　　Part II, dealing with Intermediate studies in Dynamic Evolution, covers many fascinating examples of the interaction of water molecules with other molecules, particularly proteins.

A Matter of Consequence

2.24　　　Some entities fit the description *consequential* in more ways than simply *coming into existence* as a result of the Big Bang. A simple example is the *consequential* formation of various elements within the heart of extremely hot objects such as stars and supernovae; a "recycling."

2.25　　　Further examples include the *consequential* creation of proteins within the human cell, and the *consequential* creation of the element lead resulting from radioactive decay of uranium. Of course, these are not necessarily *inevitable* consequences; e.g. not all uranium becomes lead.

The Plexus

3 THE PLEXUS

Outline

3.1   As stated earlier, the underlying principles of Dynamic Evolution begin at the sub-atomic level and extend upwards to the vast Galactic Superclusters, hence covering the "building blocks" of all the *Consequential Entities* that comprise the universe around us. Although the principles of quantum mechanics also apply to atoms and molecules, this level is nevertheless placed first. The illustration below demonstrates that, not only are these 12 levels inter-connected and exhibit close relationships based on **consistent rules**, but the Consequential Entities within each level are **linked by certain characteristics and similarities of nature, function, form**, or other facets.

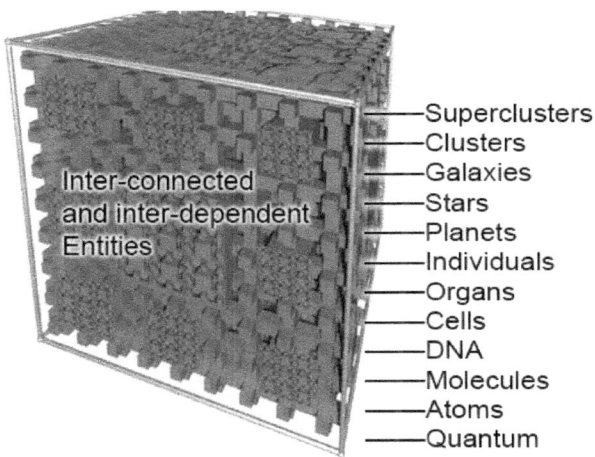

Figure 5 - A basic view of the 12 levels of The Plexus—where the fractal principles and the Consequential Entities of Dynamic Evolution live

3.2   These levels are characterized by the 12 major headings: Quantum, Atoms (which includes fundamental forces

	such as the strong nuclear force and gravity), Molecules, DNA, the Cell, Organs, Individuals, Planets, Stars, Galaxies, Galactic Clusters, and Galactic Superclusters.
3.3	The Consequential Entities are each seen within the context of one or more of these levels. Therefore the **relationships between these Entities form imaginary connecting filaments**, not unlike the plexus of the human nervous system that extend outwards in all directions. It becomes a 3D model.
3.4	Accounting for the factor of time, this could be regarded as a 4D model. The Consequential Entities are examined from the **earliest possible times up to the present day**; therefore it appears to be inadequate to make reference to a *3D* model.
3.5	By virtue of the multitude of these "connecting filaments," and the latticework picture that emerges, it is appropriate to refer to this model as **The Plexus**. This is discussed at greater length in the next chapter. However, its purpose is to enable a fuller understanding of the principles of Dynamic Evolution and the consistency and viability of those principles.

Categories of Entities

3.6	The *Consequential Entities* fall into four broad categories as well as being assigned to different levels within the Plexus.
3.7	Category 1: Failures that mean either the universe **cannot exist at all**, or conditions are inadequate to support human life.
3.8	Category 2: A "wobble" that prevents a key component from existing, that is, something that **reduces the quality of human life** to a *significant* degree.

3.9   Category 3 refers to conditions, forces, or objects of lesser importance that nevertheless enhance the quality of life and that possess special intrinsic properties that are **exceptionally difficult to attribute to the Darwinian model**.

3.10   Category 4 Entities have no known or no substantial impact on the Plexus or on other Entities if they fail or if they do not exist. However, they nevertheless have a **specific place within the Plexus** that is adequately explained by Dynamic Evolution but only fits a Darwinian explanation by means of the **Mastermind Effect** ().

Examples of Categories

3.11   An example of a Category 1 Entity is the **polarity of the water molecule**. If it was a small percentage too high or too low, life (as we know it) would not be possible anywhere in the universe.

3.12   An example of a Category 2 Entity is the **traffic control** of every human cell. The multitude of "two-way doors" that each cell possesses controls the objects that are permitted or denied access to the cell. This control is achieved by identifying either the shape, the electrical charge or the size of the object. If the **charge/shape/size of a nutritional element just didn't happen to fit or match**, then perpetual health problems would be the curse of humankind. Dynamic Evolution, therefore, has **the ability to provide the necessary links** that establish the correct relationship that both Entities require.

3.13   Category 3 Entities are numerous. Take the example of **bones** in the human body: Dynamic Evolution ensures that they are **hollow to serve as nurseries** for developing blood cells; thick and sturdy where they need to bear greater body weight; they frequently have **protrusions** that enable muscles to attach; they are also **grooved to enable nerve fibers and blood vessels** to extend smoothly. Bone is four times as strong as concrete and, pound-for-pound, is stronger than steel, due to its organic-inorganic consistency and its intrinsic structure of **nanocrystals** only **3 nanometers thick**. (Additional *Entities* and attributes of bone are covered at greater length in Part II.) And introducing the *time* element of the Plexus: In the womb, the heart is formed at the beginning of embryonic development whilst the bones do not begin to form until the fetal stage! An example of *Dynamic* principles!

3.14 An interesting example of a Category 4 Entity is the study of the tear ducts of the eye. These are accurately **placed by Dynamic Evolution at the upper, outer corner of the eye** so that cleansing fluid flows down to cover the entire surface of the eye, with the drainage system in the lower, inside corner. Examples in nature where these tear ducts have occurred in a different arrangement, for example at the bottom or side of the eye, are lacking, and therefore could not have been subject to a *natural selection* along Darwinian principles. Accordingly, there are no examples of eyes that evolved tear ducts but failed to evolve the corresponding drainage system, which would result in fluid inconveniently overflowing down the face at regular intervals, and moreover would prevent any *natural selection* from being possible.

3.15 As a curious Entity-relationship comparison (i.e. the interconnecting filaments within the *Plexus*), an involuntary **blink of the eyelid is preceded by an automatic action** that shuts down the visual section of the brain so that we do not register a "blank screen," even for a moment. Thus, the principles of Dynamic Evolution have enabled multiple facets of the human eye to emerge as closely-related, feature-rich, Entities with their properties and attributes.

Figure 6 - Fractal Electrons serve to illustrate the inter-connected principles of Dynamic Evolution as portrayed in The Plexus
(image courtesy of http://www.capturedlightning.com)

Entities and their Place in the Plexus

3.16 As mentioned (§3.1), each *Entity* may belong to one or more of the 12 levels of the Plexus. The Plexus portrays the mutually dependent **Consequential Entities** and demonstrates that they each have a symbiotic relationship with close-proximity or homogeneous *Entities* along parallel or connected latticework pathways of the **Plexus**. For example, the *Entity* water, and its various attributes, are cross-referenced directly to a very large number of other *Entities* (some of which also have their own specific properties and attributes) which have a **dependency on these Entities for their existence and continuity**.

Figure 7 - Water molecules interacting with other molecules. Bonding takes place on the blue surface in the center of the image (http://www.schrodinger.com)

3.17   As explained above, some of these objects also belong to the category *Fine-tuned Entities* which relate to the various processes, elements, and conditions that exist in the universe with **very specific parameters** (sizes, speeds, strengths, distances, etc.) that only work effectively (or at all) if those parameters are within a particular value threshold; for example, plus or minus 1% (+-1%). One example is the **strong nuclear force** which has a value of 0.007 — if it happened to vary by **only one-thousandth of a percentage point either way**, nothing of any interest would exist in the universe, let alone any form of life. Dynamic Evolution explains how this fine-tuning fits into the Plexus model. The Darwinian model does not attempt to explain this, but suffers accordingly as the principle of natural selection fails from **an inability to "select" alternative settings for this critical value**.

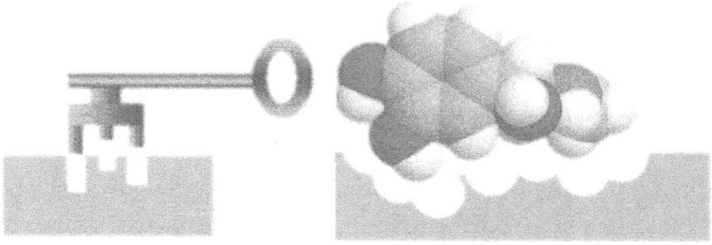

Figure 8 - A simplified (artistic) example of protein bonding using specific shapes

3.18   The above illustration provides a literal example of one of the "connection types" within the Plexus. Proteins also bond using other attributes such as electrical charge or particular oil-water constituency.

3.19   Proteins are created within the cell, and each protein has a specified shape that determines the function it will carry out within the cell. The *logical connections* that exist between proteins and other molecules are an excellent example of the Entity connections within the Plexus. However, it is easy to get confused between *literal* connections like the example of protein bonding in the picture above, and the *allegorical* connections referred to when discussing Consequential Entities within the Plexus. Remember that Entities are "all the processes, elemental objects, recombinant objects, conditions, and forces that exist and have existed in the universe" (§2.6). **The relationships between these Entities, then, is a logical connection based on physical or functional characteristics**. There are several types of connections that exist within the Plexus, and these are explained fully in Part II. The connection that the protein illustration above demonstrates is one of "form." If it were not for the shapes of the various cell-component-types, a multitude of vital functions within the cell would simply not occur. Many other Entities depend on the principle of *form*, and

Examples of Consequential Entities

3.20 There are many examples and descriptions of *Fine-tuned* Entities to be found simply by using an Internet search engine. The exponents of Darwin are aware of these lists and deal with some of them using results from various scientific tests and theories. However, we must take into account the **interlocking, interdependent, relational** nature of these Entities at all *levels* and under all *conditions;* otherwise our proverbial "mosaic" will begin to look rather murky, inconsistent, and unpleasant.

3.21 For example, in Part II, the Organs section of the *Plexus* covers the labyrinthine relationships of the different body organs, tying these in with fine-tuning at the *cellular level* of each organ and the interaction at the *molecular level* with the delicate fine-tuning of water molecules—**an inextricable linkage of Entities!**

3.22 One more example from the advanced studies in this series before we move on: According to *The Encyclopædia Britannica*:

> "Besides the blood which actually circulates in the arteries, veins and capillaries, the body possesses reserves which can be mobilized. One such is known to be located in the spleen. On the onset of hemorrhage the spleen shrinks, squeezing blood as from a sponge into the circulation."

3.23 Consequently, there is a complex link between the Plexus Entities of the brain, blood vessels, heart, and spleen that serves a vital purpose in the event of an emergency.

3.24 **There are theories that endeavor to explain some aspects of the behavior of these *Entities*** and the answers given by Darwin adherents may influence the direction that students take in their studies.

3.25 **This partly defines the purpose and nature of *Dynamic Evolution*.** It cross-references the *Consequential Entities*, along with their *properties* and *attributes*, within the Plexus at multiple levels and with many connection types. The relationships between same-level Entities is expanded upon, as well as the relationships between different levels within the Plexus.

3.26 A full understanding of these related arguments (and the *nature* of the relationships) will help the investigator to take another realistic look at the Darwinian model. The absurdly large number, represented in the **Eccentricity Diagram** (covered in Part III), is an exiguous fraction of the conceived total resulting from these *Entities* and their multi-layered Plexus relationships.

3.27 Compare this number (which is actually a three-dimensional version of the Eccentricity Diagram) with the **total number of elementary particles in the entire universe** = a relatively small 10^{80} (a number followed by *just* 80 zeroes).

3.28 Consequently, even if the processes of Darwin's natural selection could somehow deal with the order and continuity required at every level in the Plexus and at every *time* stage (assuming that the Darwin model endeavored to cover these aspects), **the universe is simply far too young by a vast order of magnitude!** Do the principles of Darwinian Evolution form any familiar congruous pathways in the above concepts? Or is this not rather a demonstration of the **superiority of Dynamic Evolution**?

3.29   The bulleted list below (beneath the table) is a very small fraction of the *Consequential Entities* list. The more extensive list, covered in Part II, contains Entities that belong to each of the following levels of the *Plexus*:

1	Quantum
2	Atoms
3	Molecules
4	DNA
5	Cells
6	Organs
7	Individuals
8	Planets
9	Stars
10	Galaxies
11	Clusters
12	Superclusters

- protons
- neutrons
- quarks
- electrons
- strong nuclear force
- weak nuclear force
- force of gravity

- electromagnetic force
- atoms
- speed of electron rotation
- angle of electron rotation
- quantum orbit of electron rotation
- electrical charge of protons & electrons
- ratio of the masses of protons & electrons
- mass of the neutrino
- decay rate of protons
- number of protons & number of electrons in the universe
- nuclear resonance for formation of carbon
- nuclear resonance for formation of oxygen
- nuclear resonance for formation of beryllium
- density of all matter after Planck time
- unique chemistry of carbon
- ratio of exotic matter mass to ordinary matter mass
- ratio of neutron mass to protein mass
- initial excess of nucleons over anti-nucleons
- ratio of matter to anti-matter in the universe
- polarity of water molecule
- thermal properties of water molecule
- capacity of water to dissolve (alcahest)
- low chemical-reactivity of water
- low viscosity of liquid water
- high viscosity of water ice
- latent heat of freezing water
- density of water
- alteration in the "law" of water contraction just before freezing
- bonding of more than 40 elements for life
- speed of light
- explosive force of the big bang
- amount of matter that existed at time of the big bang
- rate of expansion of the universe at all stages
- size of relativistic time dilation factor
- "uncertainty magnitude" in Heisenberg uncertainty principle
- distance of moon from the earth
- distance of earth from the sun
- size of moon

- size of earth
- tilt of earth
- average distance between stars
- average distance between galaxies
- average distance between galactic clusters
- average distance between galactic superclusters
- position of sun in relation to both center and edge of Milky Way
- number of dimensions in the universe
- level of dynamic entropy at all stages in formation of universe
- strength of force of gravity at all *time* stages in formation of universe
- strength of force of gravity at all *size* stages from atoms to superclusters
- frequency of occurrence of supernovae eruptions
- distribution of supernovae eruptions
- quantity of white dwarf binaries
- timing of development of white dwarf binaries during formation of universe

3.30   The above constitutes a (very) partial list of the universal forces, objects, and conditions that make up the *Consequential Entities*. Most of the above items only relate to the first one or two layers of the Plexus.

3.31   Some of these Entities are so finely balanced, so fine-tuned, that a "wobble" of the tiniest of fractions would upset one or more of the components of the Plexus with the resulting cascade failure of the entire model (i.e. either the catastrophic failure of a key component essential to mankind's life continuity, or the impossibility of the universe or of human life ever existing at all).

3.32   The mental image that the above explanation evokes is a very **poor imitation of the real *Plexus***, which contains substantially more *Entities* multiplied by many orders of magnitude, and many more connecting filaments than any artist could effectively illustrate.

Latticework — the Eye-Candy of the Plexus

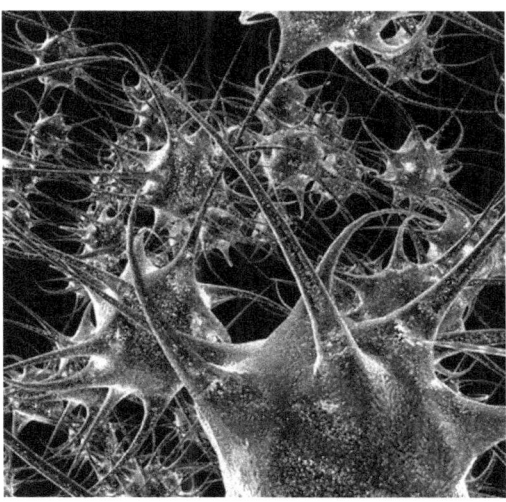

3.33   The *Plexus* is so-named because of the neural network-type "latticework" connections between the *Entities* (and their *properties* and *attributes*) at all levels within this model that demonstrate how tiny a fraction of the Plexus the numbers illustrated in the Eccentricity Diagram (discussed in Part III) are.

3.34   The model's components are also characterized by a "syzygy" (as in an alignment of planets) due to the congruity and order in aspects of the *Entities* and levels of the Plexus (one example is the **periodic table** of elements, another is the use of **Fibonacci numbers** and the **Golden Ratio** in the plant (the logarithmic spirals of shells, buds in flowers, the bracts of pinecones) and animal Kingdoms (from the bone segments in the human hand to the features of the human face, and from microbes to the largest creatures on earth) — Category 1 and Category 4 *Consequential Entities* respectively). But the key qualities of the Plexus are the **inter-connectivity, filial**

relationships, and other attributes manifest with respect to the Entities in the model:

3.35　Just as the 100 billion neurons in the human brain can be connected via axons and dendrites to anything up to 10,000 other neurons each (considerably more for certain alternative cell types), so too the *Consequential Entities* of the *Plexus* are characterized by **a seeming infinity of closely-related connections**.

3.36　Even the human brain itself provides a *literal* example of this concept: Each neuron maintains a voltage gradient across its membrane, due to differences established by cellular metabolism in ions of sodium, potassium, chloride and calcium within the cell, each of which has a different charge. If the voltage changes, then **information transmission/reception will be degraded or ineffective**. The process of communication between neurons is highly complex, involving chemical and electrical connections at the molecular level. Hence, the Entity-relationships **reach multiple levels** within the Plexus with innumerable mutually-dependent conditions.

3.37　The model assembles these *Entities* into 12 categories, or levels, related to objects of varying sizes; starting from the **quantum level** at the lowest and finishing at the highest level ... currently this is believed to be 'galactic superclusters' (although a 13th level of *'super* superclusters' may be indicated).

3.38　The complete *Plexus* demonstrates the **relation-ships between all of the elements** within the model by cross-referencing to (1) other *Entities* within the same level, (2) to *Entities* that belong to other levels within the *Plexus*, (3) to explanatory text in other articles within this work, and (4) to the work of science writers both in books and on the Web.

3.39   Note that, as discussed earlier, only some of the *Consequential Entities* result in catastrophic failure of the model if a "wobble" occurs (Category 1 entities), that is, a failure that means either the **universe cannot exist** at all, or conditions are inadequate to support human life.

3.40   Category 2 indicates a "wobble" that prevents a key component from existing; that is, something that **reduces the quality of human life** to a *significant* degree.

3.41   Category 3 refers to conditions of lesser importance that nevertheless **enhance the quality of life** and that possess special intrinsic properties that are **exceptionally difficult or impossible to attribute to the Darwinian model**.

3.42   Category 4 *Entities* have no known or no substantial impact on the *Plexus* or on other *Entities* if they fail or if they do not exist. However, they nevertheless have **a specific place within the Plexus** that is adequately explained by Dynamic Evolution.

Compartmentalizing the Consequential Entities

3.43   In each of the tables below, the following legend applies:

– generally represents smaller, or slower, or less dense, or younger properties

+ generally represents larger, or faster, or more dense, or older properties

If no + or – is given, the *Entity* has no specific fine-tuning properties

The 'CAT' column gives the Category number of the *Entity* (1-4).

CONSEQUENTIAL ENTITIES	
ATOMS & BASIC CONDITIONS OF MATTER	
ENTITY NAME & ATTRIBUTES	CAT
Strong nuclear force	1
+ no hydrogen would form; atomic nuclei for life-essential elements would be unstable; life would not be possible	
− no elements heavier than hydrogen would exist: life would not be possible	
Weak nuclear force constant	1
+ too much hydrogen would convert to helium; stars would convert too much matter into heavy elements; life would not be possible	
− too little helium would be produced; stars would convert too little matter into heavy elements; life would not be possible	
Strength of force of gravity at all *time* stages of universe	1
+ stars would be too hot and would burn too rapidly and too unevenly	
− stars would be too cool for nuclear fusion; elements needed for life would never exist	
Strength of force of gravity at all *size* stages of universe	1
+ gravity would overwhelm universal expansion and fatal contraction would occur	
− gaseous nebula would never succeed in forming stars	
Electromagnetic force	1

CONSEQUENTIAL ENTITIES	
ATOMS & BASIC CONDITIONS OF MATTER	
ENTITY NAME & ATTRIBUTES	CAT
+ chemical bonding would not occur; essential elements would be unstable	
− chemical bonding would mean life would not be possible	
Ratio of electromagnetic force to gravitational force	1
+ all stars would be at least 1.4 times mass of the sun; life-cycle of star would be too brief to support life	
− all stars would be at least 0.8 times mass of the sun, making them incapable of producing heavy elements	
Mass of the neutrino	1
+ If neutrinos have even a small amount of mass, their high density throughout the universe would increase the *Omega* value (the mass in the universe) causing its eventual collapse	
− If *Omega* (the mass in the universe) is infinitesimally less than 1, it cannot prevent the universe from expanding forever	
+ galaxy clusters and galaxies would be too dense	
− galaxy clusters, galaxies, and stars would not form	
The lambda particle	2
+ If lambda ("vacuum energy" or "quintessence") is non-zero, universal expansion may actually be accelerating	
− If lambda is zero, the universe may collapse	

CONSEQUENTIAL ENTITIES	
ATOMS & BASIC CONDITIONS OF MATTER	
ENTITY NAME & ATTRIBUTES	CAT
Ratio of electron to proton mass	2
+ chemical bonding would mean life would not be possible	
− [as above]	
Ratio of number of protons to number of electrons	1
+ electromagnetic force would be too great for gravity, preventing formation of galaxies, stars, planets	
− [as above]	
Expansion rate of the universe	1
+ no galaxies would exist	
− universe would collapse	
Entropy level of the universe	1
+ stars would not exist within proto-galaxies	
− proto-galaxies would not exist	
Mass density of the universe	1
+ excess of deuterium would mean stars would burn-out too rapidly for life to exist	
− insufficient helium would mean shortage of heavier elements	
Velocity of light	3
+ stars would be too bright	

CONSEQUENTIAL ENTITIES	
ATOMS & BASIC CONDITIONS OF MATTER	
ENTITY NAME & ATTRIBUTES	CAT
− stars would be too dark	
Age of the universe	1
+ no stars sufficiently stable would exist in required locations of the galaxy	
− stable stars would not have formed	
Initial uniformity of radiation	1
+ *if more uniform*: stars, star clusters, galaxies and galactic clusters would not exist	
− *if less uniform*: universe would quickly have become black holes and be predominantly empty	
Distance of moon from the earth	2
+ If much closer, tidal waves would be 1,000 times greater than they are today	
− If much further away, earth's day would be only 8 hours; winds & hurricanes would be considerably greater; oceans would not be chemical-rich & therefore inadequate for life to begin	
Distance of earth from the sun	1
+ Freezing temperatures would not permit life to survive, or to begin to develop	
− Heat would scorch the atmosphere, as well as land; oceans would be evaporated, and therefore no chemicals required for life-synthesis would exist	
Size of moon	2
+ Sun's gravitational effect on moon (currently 2x that of the earth) would be greater, causing severe	

CONSEQUENTIAL ENTITIES		
ATOMS & BASIC CONDITIONS OF MATTER		
ENTITY NAME & ATTRIBUTES		CAT
	irregularities in moon's orbit with consequences similar to those explained for '**Distance of moon from the earth**'	
−	The moon would have less mass and would therefore draw rapidly closer and closer to the earth; day would be only about 15 hours long. Less scattered sunlight; greater seasonal fluctuations	
Average distance between stars		1
+	heavy element density would be too sparse for rocky planets to form	
−	planetary orbits would be too unstable for life	
Average distance between galaxies		1
+	lack of material for star formation	
−	gravitational effects would destabilize the sun's orbit	
Density of galactic cluster		1
+	galaxy collisions and mergers would destabilize the sun's orbit	
−	lack of material for star formation	
Fine structure constant		1
+	stars would have significantly less mass than the sun; matter would be unstable in large magnetic fields	
−	all stars would have significantly greater mass than the sun	

CONSEQUENTIAL ENTITIES	
ATOMS & BASIC CONDITIONS OF MATTER	
ENTITY NAME & ATTRIBUTES	CAT
Decay rate of protons	1
+ radiation would prevent the existence of life	
− universe would contain insufficient matter for life	
Ratio of neutron mass to proton mass	1
+ neutron decay would yield too few neutrons for the formation of many life-essential elements	
− neutron decay would produce so many neutrons that all stars would collapse into neutron stars or black holes	
Initial excess of nucleons over anti-nucleons	1
+ radiation would prohibit planet formation	
− available of matter would be insufficient for galaxy or star formation	
Supernovae eruptions	1
+ *if too distant, too infrequent, or too soon*: heavy elements would be too sparse for rocky planets to form	
− *if too close, too frequent, or too late*: radiation would exterminate life on the planet	
White dwarf binaries	1
− *if too few*: insufficient fluorine would exist for life chemistry	
+ *if too many*: planetary orbits would be too unstable for life	

CONSEQUENTIAL ENTITIES			
ATOMS & BASIC CONDITIONS OF MATTER			
ENTITY NAME & ATTRIBUTES			CAT
	−	*if formed too soon*: insufficient fluorine production	
	+	*if formed too late*: fluorine would arrive too late for life chemistry	
Ratio of exotic matter mass to ordinary matter mass			1
	+	universe would collapse before solar-type stars could form	
	−	no galaxies would form	
Number of dimensions in early universe			1
	+	quantum mechanics, gravity, and relativity could not coexist; thus, life would be impossible	
	−	same result	
Number of dimensions in present universe			1
	−	same result	
	+	electron, planet, and star orbits would be unstable	
Big bang ripples			1
	+	galaxies/galaxy clusters would be too dense for life; black holes would dominate; universe would collapse before life-site could form	
	−	galaxies would not form; universe would expand too rapidly	
Cosmological constant			1

CONSEQUENTIAL ENTITIES	
ATOMS & BASIC CONDITIONS OF MATTER	
ENTITY NAME & ATTRIBUTES	CAT
+ universe would expand too quickly to form solar-type stars	

CONSEQUENTIAL ENTITIES	
MOLECULES	
ENTITY NAME & ATTRIBUTES	CAT
Polarity of the water molecule	1
+ heat of fusion and vaporization would be too high for life	
− heat of fusion and vaporization would be too low for life; liquid water would not work as a solvent for life chemistry; ice would not float, runaway freeze-up would occur	
Size of the relativistic dilation factor	2
+ certain life-essential chemical reactions would not function properly	
− same result	
Uncertainty magnitude in the Heisenberg uncertainty principle	2
+ oxygen transport to body cells would be too great and life-essential elements would be unstable	
− oxygen transport to body cells would be too small and life-essential elements would be unstable	
12C to 16O nuclear energy level ratio	1
+ insufficient oxygen for life	

CONSEQUENTIAL ENTITIES	
MOLECULES	
ENTITY NAME & ATTRIBUTES	CAT
− insufficient carbon for life	
Ground state energy level for 4He	1
+ insufficient carbon and oxygen for life	
− same as above	
Decay rate of 8Be	1
+ no element heavier than beryllium would form; no life chemistry	
− heavy element fusion would generate catastrophic explosions in stars	

3.44   The above tables contain many references to advanced principles that admittedly require much explanation. As stated earlier, this work is not intended as a textbook to teach every associated scientific principle. Nevertheless, later studies provide a systematic approach to these principles as they relate to Dynamic Evolution, and key concepts are dealt with in greater depth.

3.45   *The Plexus* (illustrated earlier) is a key teaching tool of Dynamic Evolution. The purpose of this tool is to demonstrate key concepts of the study in diagrammatic form. Entities that are inextricably linked together by logical argumentation or by a consistency of design or by other connections, are given their specific place within the model. The reason for referring to this as a *cubic* model is that Entities have relationships within the same level (*horizontal*—either in parallel or by secondary or remote connections) *as well as* with Entities on other levels (*vertical*). Examples of these arrangements are provided below.

3.46	The name **Plexus** was chosen because of the neural network-type "latticework" connections between the Entities at all levels within the Plexus that demonstrate the **consistency of the attributes** peculiar to the 3-dimensional model (4D if the attribute of time is included).

The Plexus — Exquisite Fractal Reticulation

3.47	The *Oxford English Dictionary* defines **reticulation** as "a pattern or arrangement of interlacing lines resembling a net." As for **fractal**, the many relationship types between Entities and their properties manifest many repeating patterns, like the elegant patterns seen in mathematically attractive fractal illustrations. These are dealt with at length in later studies.

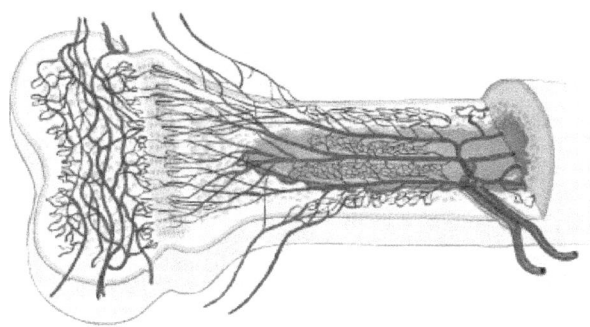

Figure 9 - Blood vessels extend for maximum efficiency and branch out according to Dynamic principles (a fractal effect)

3.48	Fractals occur in various forms within the Plexus; one example can be illustrated by trees which are fractal in form: a tree is characterized by limbs that extend outwards at restricted angles, and the branches extend in a similar fashion from the limbs, and so on with the smaller branches, then twigs, then stems for leaves. **In these cases, there is a similarity of form—which accompanies the iterative decrease in size**—that gives the tree its fractal appearance. Likewise, the fractal reticulation of the principles of **Dynamic Evolution** manifest in the Plexus—Entities, Properties, Attributes, connection types, and the more advanced aspects discussed in further studies—have a **similarity of form, behavior, and appearance from the point of view of their mathematical and functional properties**.

Trees manifest a fractal form

Figure 10 - Fractal patterning of Sunflowers

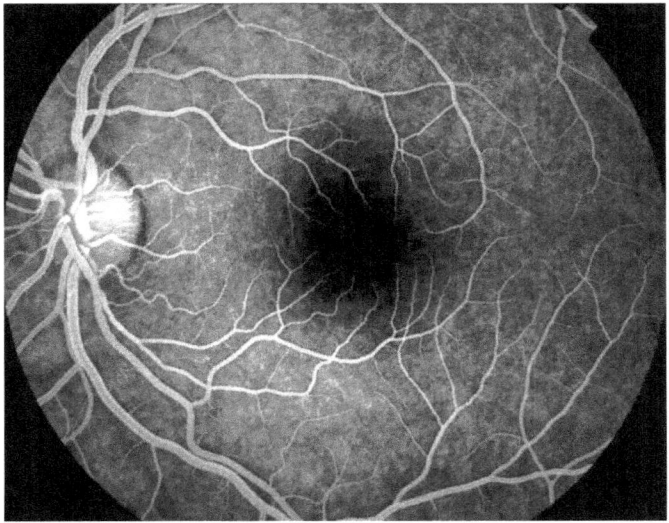

Figure 11 - Fractal branching within the eye

Fractal arrangements occur throughout nature, manifesting a similarity of form with each iterative decrease in size

An Outline
of
Dynamic Evolution

4   An Outline of Dynamic Evolution

Outline

4.1 Dynamic Evolution provides the argumentation for the progressive development of the universe and of life which manifest coherent and connected laws, parameters, and conditions that govern **how, where, and when a given element, universal force, or other Entity (see §2.6, *Consequential Entities*) finds its place within the *Plexus*** (see previous chapter). This work is not concerned with the implications inherent in these principles; it is only concerned with the evidential principles themselves.

Figure 12 - Illustrating the compartmentalizing of the fractal principles of Dynamic Evolution (Picture by http://paulbourke.net/fractals/menger_sl)

4.2 The operative word in the above paragraph is probably "how." Each *Entity* has a **specific place** within the *Plexus*. This has an impact on the closely-related, interdependent objects within the Plexus in terms of the **logical connections intrinsic to the model's stability**.

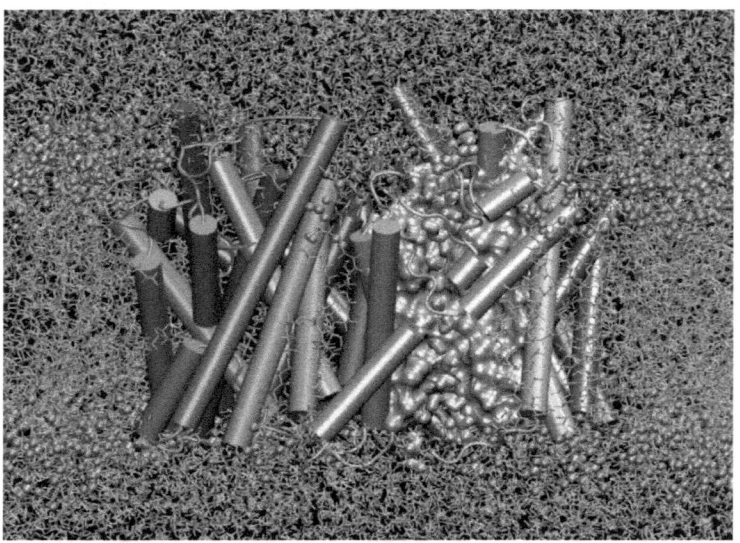

Figure 13 - Artist's impression of aquaporins controlling transport of water

4.3 For example, the multiple fine-tuned attributes of the water molecule Entity cross-reference to fine-tuned functions within the human cell, both in terms of **time** (e.g. human conception as well as embryonic and fetal development) and **critical functional relationships** such as temperature control, aquaporin organization for transporting bidirectional water molecules (controlling up to a billion per second, per aquaporin, per cell—in the kidneys alone these proteins are in charge of re-absorption of about 40 to 50 gallons of water every day), and the diffusion and osmosis of other substances; and in turn establish further plexus relationships along these related pathways within the model, to body organs such as the liver (see 'Multi-functional Entities' below), the stomach, the kidneys, and to body systems such as the pulmonary, systemic, lymphatic, and digestive: the resulting illustration of the connective-lattice being **too complex for any artist to construct effectively**.

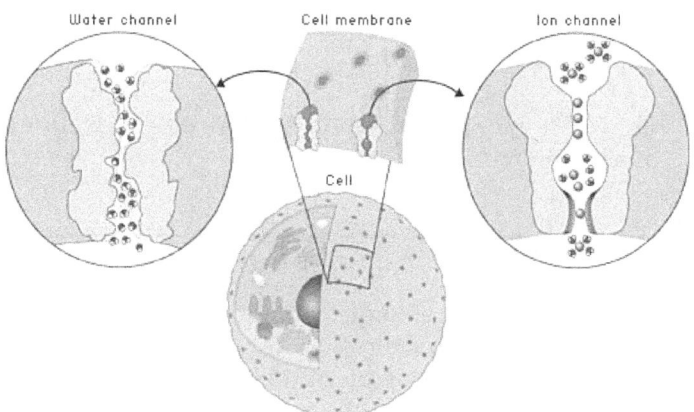

Figure 14 - Controlling water transportation involves chemical and electrical signals within the cell

4.4   Apart from the membrane of human cells, these complex proteins, aquaporins, have been identified in many other organisms such as bacteria, fungi, plants, mammals, etc. Thus, the cross-linked connections of the Plexus and the principles of Dynamic Evolution **branch out in fractal fashion to many other areas of science** (as illustrated by the connections to Entities in the many levels in the Plexus).

4.5   However, we are only reviewing the fundamentals of Dynamic Evolution at this stage. If time was in greater abundance and 'work space' was not at a premium, we could discuss the interactions between the water molecule's various *attributes* and *properties* and a selection of thousands of other functional Entities (and their attributes) within the model in, for example, the **Molecules** level of the *Plexus*. We would then **continue the latticework matrix** along each of the channels in a logical progression up through the levels of the Plexus for every level and every connection **where at least one dependency existed**.

4.6   The above explanation constitutes an overview of the basics of Dynamic Evolution.

Dynamic vs Darwinian Evolution

4.7   Contrast the consistency, density, and immense scope of the Plexus as conceived so far in the above description, with some of the explanations proffered by Darwinian Evolutionists when they explain how *independent* yet *inter-dependent* Entities arrived by the Darwinian model. Note the following expressions:

"bacteria … **made a choice** to use this substrate"

"…the beak grew longer **in order to deal with** the tougher seeds."

"…fruit flies **needed to change** into different kinds in order to live on the different islands."

"…the Skunk **wanted a defense** mechanism…"

"…a better defense mechanism was required … evolution **provided the answer**…"

"nature **designed** DNA…"

"there is an evolutionary **purpose** to feeling really sad"

"…the function that the brain **evolved to perform**."

4.8   There is no need to show further examples; they are replete in numerous works promoting Darwinian Evolution. Some Darwin proponents excuse the proliferation of these remarks by explaining that they are merely expressions that do not *effectively* teach the point under discussion, or that only "lesser" scientists are responsible for such faux pas. However, even the highly respected professor, the late Carl Sagan, on his TV show

Cosmos some decades ago, regarding the *need* for greater informational storage capacity than DNA, said that, **after this need arose, "we invented the brain!"** The temptation to attribute Dynamic principles to evolution nevertheless persists.

4.9 What is interesting at this point is a comparison of the above **anthropomorphic Freudian slips** with the principles of Dynamic Evolution. If a particular functionality is required for survival, a variety of explanations are offered for the existence of principles that explain how Darwin's evolution "accomplished the task." **The seduction of the principles of Dynamic Evolution seem to be unavoidable:**

> A body **system** was needed, therefore...
> an **organ** (one or more) was needed, therefore ...
> a **cell-type** was needed, therefore ...
> a special **protein-type** was needed, therefore ...
> a specially-formed bonding **molecule** was needed, therefore ...
> a particular atomic **element** was needed...

4.10 Let's take a contrived example for the purpose of illustration: Construction workers at a manufacturer that produces sophisticated spacecraft... say, a Nasa / IBM / Boeing kind-of conglomerate. There are multiple departments at this plant, which manufactures all of its own components; from the nuts and bolts to the fuselage, from the micro-fibers to the seats, from the Silicon chips to the million-line-code software programs. You approach a number of employees and ask them what work they do. They show you the component that they make, and they're very skilled. You then ask them what the component does. But they have no idea what part their component plays in the finished product...

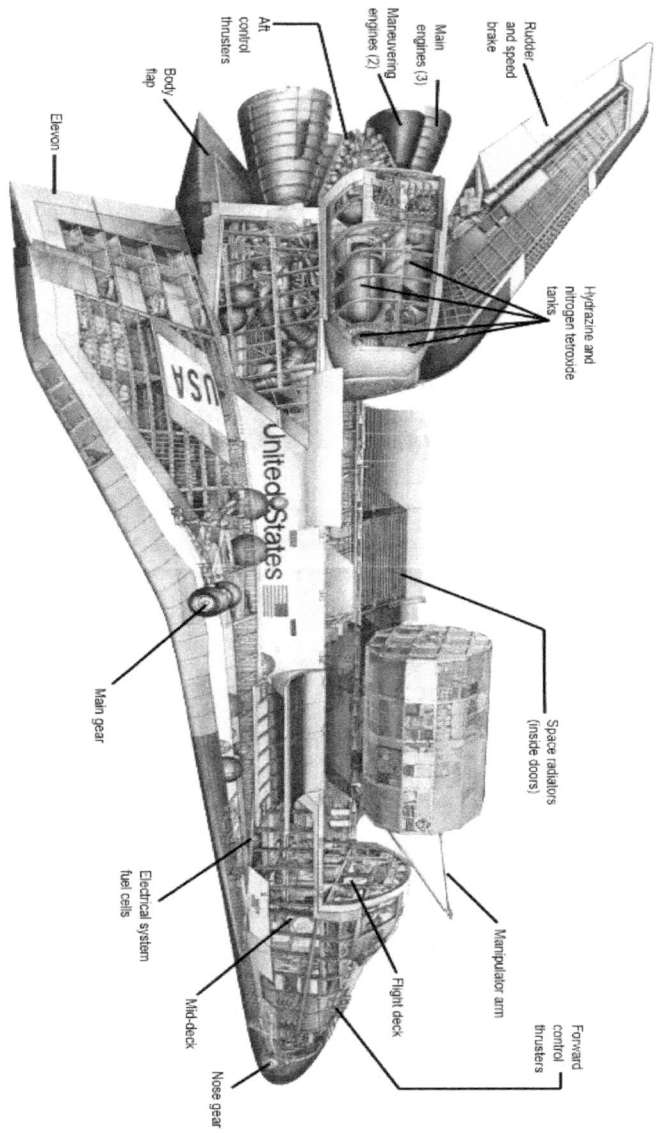

Figure 15 - Space Shuttle Schematic (courtesy of Nasa)

4.11   How do they know how to make the component? They follow a *plan* (which corresponds to a set of *principles* within Dynamic Evolution). A supervisor takes their finished product and assembles it, along with several hundred other components made by equally institutionalized employees, each beavering away in his own specially fitted area of the factory floor. You question the supervisor, but **he still doesn't know what part his assembled component plays** in the finished product [he simply observes the pathway within the "latticework Plexus"]. You ask him about the components that his part consists of, but he has no idea how they work.

4.12   This occurs throughout five or six levels until a specialist team take each of the final assembled parts and complete the spacecraft construction. You ask each of them about the components the craft is composed of, but **they have no idea how the individual components work**, they only follow the assembly plan and instructions they have been given. And of course this spacecraft has all the sophistication that human endeavors can muster.

4.13   We are now building a picture of a basic example of Dynamic Evolution. For **we have not yet advanced beyond the fundamentals**; the scale of the Plexus, from its lowest level to its highest, means that it contains intrinsic connections **for each of the contextual Entities**. Therefore:

- What use is the plastic coating around the electrical wire unless it is graded appropriately, precision molded, and correctly fitted for insulation?

- What use is the electrical wire unless it is connected to the proper channels and interfaces?

- What use are the conduits and channels unless they lead to the power source?

- What use is the power source unless it is properly configured, active, and provided with the means to keep it topped-up with the "juice" it requires to function?

4.14 And, of course, these questions cover all of the components in this large construction project from the very tiny to the very large, involving electrical distribution, fluid transport, air channeling, communications, comfortable and ergonomic furniture, etc. **The complete plexus arrangement of component dependencies is vital for the finished product**, just as the intrinsic values and members of the Plexus are vital for the existence and continuity of the cubic model.

4.15 However, how does this apply to another construction?

- What use or purpose is this particular group of atoms **unless they combine** to make a molecule? ...

- What use or purpose is this molecule unless it has the **precise balance of properties** that enable it to interact and bond with other (appropriate) molecules? ...

- What use or purpose is this single molecule **unless it combines** with others to make, for example, DNA? ...

- What use or purpose is DNA unless it can contribute to the **production of a complete cell**? ...

- What use or purpose is this single cell unless it can join with other cells to **make an organ** (with due apologies to single-celled organisms)? ...

- And of course, what good are all the individual parts unless we have a **complete body**? ...

4.16 The "contextual Entities" are linked by many unbreakable connections, not just those covered in the bulleted list above; and the sheer volume of these resulting connections is becoming more apparent. But now **additional factors are introduced**; new segments along the horizontal levels of the 4D Plexus. One such factor is **the use of multi-functional components which have dependency links** to numerous lines of the complex latticework. The proclivity for Dynamic Evolution in the completion of the above spacecraft is already becoming irresistible. Let us add each of these new factors in turn.

4.17 Returning to the 'engraved rock' illustration discussed in Part III, the new picture (adding the above argumentation into the analogy) **now becomes a city-sized piece of ground on which are millions of carefully placed rocks** neatly aligned, each bearing its own unique, verbose inscription.

Multi-functional Entities of the Plexus

4.18 The development of a large number of Entities fits into the category of *multi-functional*. Without d*ynamic* development, the Darwinian exponent is hard-pressed to explain how a new function, by *natural selection*, served a new purpose for an organism which was already equipped to serve another purpose (or purposes).

4.19 An example of the *Dynamic* element of these principles can be seen from the following account: Late in the 20th century, a fossil moth egg was found in 75-million-year-old sediments in Massachusetts. The egg is positively assigned to the moth family *Noctuidae* and extends the fossil record of this family back into the Cretaceous. Is this significant for Dynamic Evolution? It turns out that *Noctuidae* family moths have **special organs for detecting the ultrasonic cries of insect-hunting bats**. The fossil record of the bats, however, only goes back to the early Eocene, perhaps 20 million years after the *Noctuidae* moths. Since no other insect predators like bats existed, either the moths developed these special organs in anticipation of the bats, or we have **another example of Dynamic Evolution**!

Multi-functional DNA

4.20	One of the components of our proverbial spacecraft is analogous to DNA. Another purpose of DNA is the construction of machines for use within the cell to carry out a large variety of operations fulfilling essential duties. These duties or tasks can, in themselves, be **linked to other components** within the Plexus both **horizontally** (e.g. interactions with other molecules) and **vertically** (e.g. manufacturing parts that other body organs need).

Multi-functional Endoplasmic Reticulum

4.21	A remarkable configuration within the body of cells is the endoplasmic reticulum. This organelle performs many functions: protein synthesis, translocation across the membrane, integration into the membrane, protein folding, post-translational modification, glycosylation, synthesis of phospholipids, and the regulation of calcium homeostasis. No small accomplishment for this labyrinthine structure.

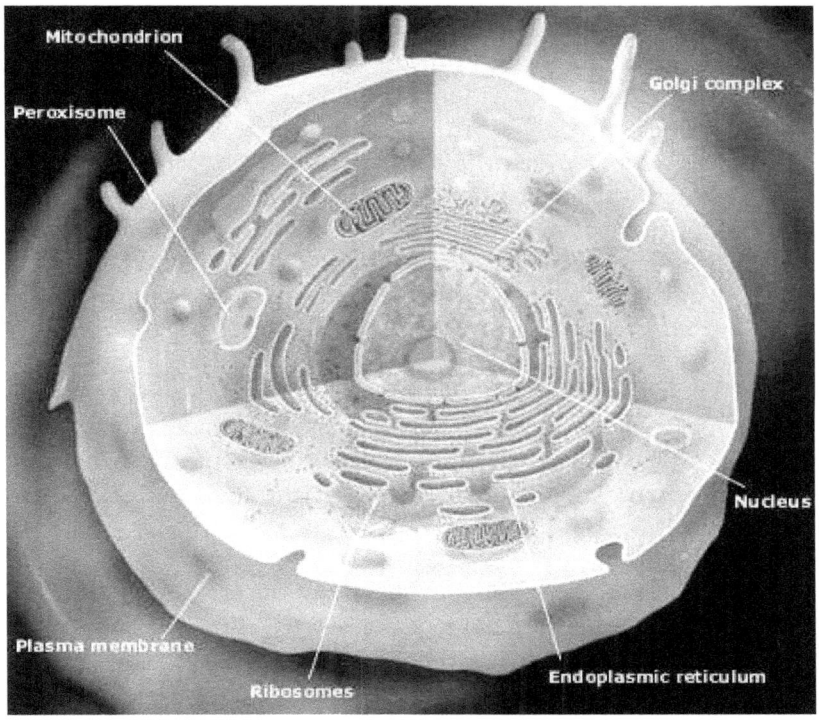

Figure 16 - Artistic depiction of the cell, showing the labyrinthine arrangement of the endoplasmic reticulum serving multiple functions within the cell body

Multi-functional Glial Cells

4.22 An example of a real (as opposed to analogous) multi-functional Entity is the **Radial Glial** cell of the human central nervous system. Although these are specialized cells vital for the developing nervous system, **they are now known to perform multiple tasks**. They are characterized by long radial processes (not unlike the Entity connectivity within the Plexus itself) that facilitate the guidance of radial migration of newborn neurons from the ventricular zone (the channels of the brain that contain fluid) to the mantle (or pallium) regions, which are the protective layers of the brain analogous to the mantle of the earth. (Incidentally, some brain cell development occurs in what scientists call "inside-out" fashion, where the deeper *dependent* cells are formed before the cells on which they depend—like watching a tree grow, but the leaves form first, followed by the stems then the branches then the limbs then the trunk! This and similar examples from different levels of the Plexus will be dealt with in more advanced articles in this series.)

4.23   The ongoing study of the human brain ensures that new levels of complexity continue to be discovered and form the basis of yet further examination. As Professor Sir Robin Murray, one of the United Kingdom's leading psychiatrists, said:

> "We won't be able to understand the brain. It is the most complex thing in the universe."

4.24   The late Dr. Isaac Asimov once said:

> "The human brain ... is the most complicated organization of matter that we know."

4.25   And as physicist Sir Roger Penrose said:

> "Consider the human brain. If you look at the entire physical cosmos, our brains are a tiny, tiny part of it. But they're the most perfectly organized part. Compared to the complexity of a brain, a galaxy is just an inert lump."

4.26   Although this work may take a little exception to the latter opinion of a galaxy being likened to "just an inert lump," nevertheless the comments regarding the human brain find agreement with Dynamic principles.

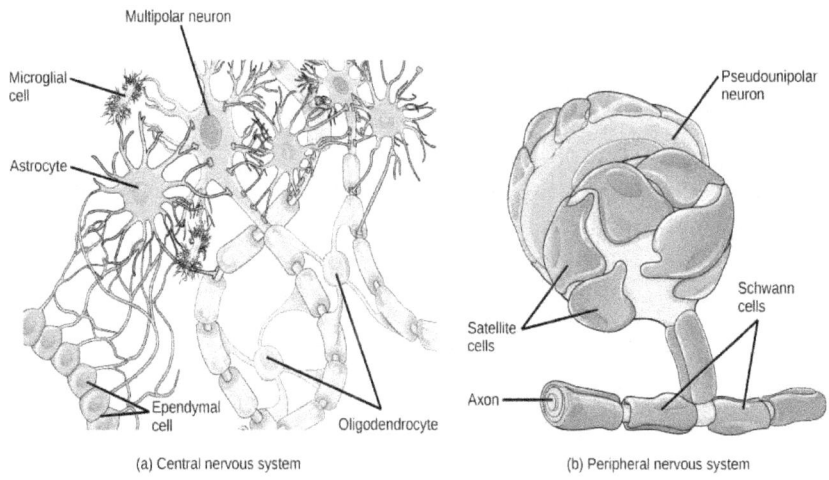

Figure 17 - Example of Radial Glial cells (courtesy Openstax College)

4.27 Recent data indicate further important roles for the brain's Radial Glial cells as **ubiquitous precursors** that generate neurons and glia, provide **maintenance tasks** for damaged neuronal connections, and also serve as **key elements in patterning** and region-specific **differentiation** of the central nervous system. They also at times perform the same **signaling functions** as neurons. Consequently, these multi-functional cells are very much involved in most aspects of brain development.

4.28 The above "roles" "functions" and "tasks" are examples of **Entity** *attributes* (covered more fully later).

Multi-functional Astrocyte Cells

4.29 Astrocytes are specialized cells found throughout the entire central nervous system; they are (for the most part) star-shaped glial cells (the name derives from the Greek words αστερ for star and κυτος for container, or cell). They are specialized glial cells that somehow **outnumber**

neurons by more than 5 to 1. Astrocytes are the most abundant cell type in the central nervous system, and for good reason; they perform many complex functions, including:

- assisting in the initial development of the nervous system (including defining the brain's overall architecture)...
- maintaining brain homeostasis (i.e. the brain's metabolic equilibrium)...
- controlling concentrations of ions, neurotransmitters, and metabolites (chemical compounds produced as a result of metabolism or metabolic reactions)...
- regulating water movements...
- removing excess (harmful) glutamate...
- supplying glutamine to maintain glutamatergic neurotransmission (a glutamate–glutamine shuttle service)...
- facilitating nerve signal transmission...
- restricting which substances can enter the brain by providing biochemical support of the endothelial cells (the thin layer of cells that line the interior surface of blood vessels and lymphatic vessels) that form the blood-brain barrier...
- controlling local blood-flow...
- delivering nutrients to nervous system tissue...
- and helping repair the spinal cord after injury.

4.30   According to the latest research by neuroscientists, the above list is by no means complete. And, as remarked earlier, what is the supporter of Darwin to make of this list? Given the closely-related Dynamic studies of other components of the central nervous system, and their own levels of complexity, how is the Darwinian system to explain the converging, mutually-dependent, yet disparate functions of these systems?

The Multi-functional Liver

4.31　　The liver is very much **a multi-functional organ**. It performs, among other things, the following functions:

    (1) It converts food into energy,

    (2) makes extensive use of water to cleanse the body from pollutants such as alcohol, drugs, poisons, and airborne mephitic chemicals;

    (3) it uses water again when manufacturing bile, even increasing production in the event of gall bladder failure; it stores

    (4) vitamins

    (5) minerals, and

    (6) energy for use by other cells and organs of the body when needed.

Figure 18 - The multi-purpose liver (illustrated by Gerry Fey)

4.32 More advanced articles in this series deal with the multilayered **dependency relationships that the complexity of this organ contrives** in the Plexus at the **atom**, **molecule**, **DNA**, and **organ** layers.

Multi-purpose Neurovascular Development

4.33 A further example involves independent neurovascular growth; that is, **nerve fibers** and **blood vessels** (both being multi-attribute Entities) during embryonic development. Growth cones of nerves and endothelial cells of blood vessels are closely analogous in the way they extend and branch out, and **they both perform similar tasks during the early development of a limb or organ**. Both must invade the mesenchyme, the plexus of embryonic connective tissue in the mesoderm (one of the three primary cellular layers of the developing embryo), which form the connective tissues of the blood and lymphatic vessels to produce complex networks of nerves and vessels. Both Entities (blood vessels and nerve fibers) must extend into regions of the developing limb such as muscles and **both form dense subcutaneous plexuses at *precisely* the same depth**. Also, adult tissues show many examples of neurovascular bundles in which nerves and blood vessels give evidence of having **developed in close parallel and have branched out in filial fashion** in a manner that well illustrates the "dependency links to multiple lines of the complex latticework" described above (§4.16).

4.34 The embryo continues to grow in size as nerve fiber and blood vessel growth continues; however, the density of both remains the same throughout the growth period until near the end of development. Moreover, during the early stages of neurovascular growth, scientists see no obvious signs of symbiotic (mutually dependent) development. They do observe, though, that blood vessels tend to be in place generally before nerve fibers. However, during the mature stages of neurovascular branching, nerve fiber and blood vessel growth is closely correlated, so much so that nerve fibers are either **very closely paralleled** by blood vessels to within 10 micron distance, or up to 4 blood vessels form a **tight sheath around the nerve fibers ensuring an adequate supply of blood** at this microscopic level. Once tissue growth is complete for the organ or limb and branching has reached its maximum twig level, neurovascular growth ceases.

4.35   The behavior of these systems throughout embryonic and fetal development is a classic example of **Dynamic Evolution**. How does each filament of fiber or blood vessel branch out (for example, cascading to twig level at appropriate density nearing the end-points) and determine its direction? How do they determine when to stop branching out? How does the branching density respond when organ or limb growth has stopped? Why does branching density reach "twig" level near the completion of branching at the perimeter of every organ or limb? Only when the tissue and organ growth is complete does the intended pattern become obvious. **The seduction of the principles of Dynamic Evolution is compelling**.

4.36   Considering the quantity of both blood vessels and nerve fibers that migrate to the nethermost regions of the human body (in fact, in one body alone, placed end-to-end these Entities would *each* stretch about **four-times around the earth**—not counting the more prolific lemniscus of the brain (the pathways that carry sensory information), which have more than the entire body combined—and neurosurgeons estimate that blood vessel quantities probably match those of nerve fibers, despite the fact that the *average* length of capillaries in the human body is only about half a millimeter!) it is easy to picture the *new* density of **The Plexus** as it now appears at this stage of the study.

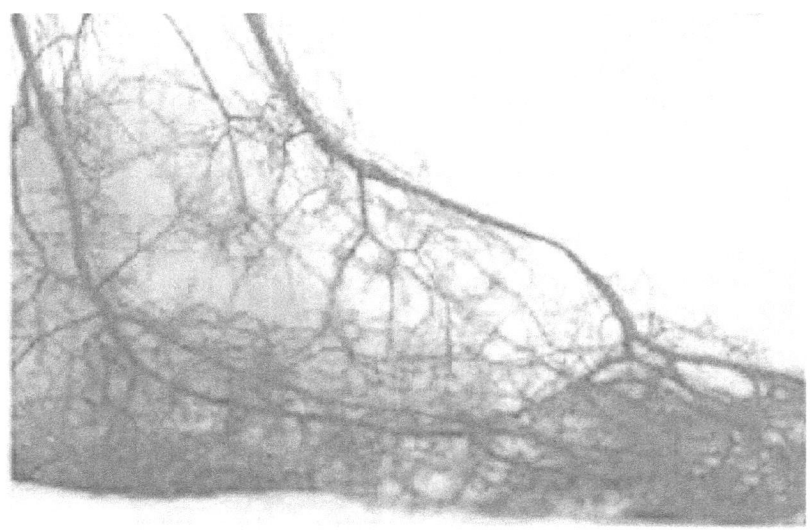

Figure 19 - Fractal branching of blood vessels in the foot
(picture courtesy of Scielo, Chile)

4.37 The above image shows the visible fractal branching of the blood vessels; lesser capillary branching is too small for the unaided eye to observe.

Peerless Attributes of Plexus Entities

4.38 An additional study in the more advanced stages of this work covers the many dependency connections of the eye (§7.21) (which Darwin predicted would be a cause of problems for his theory) and of the **nerve signal traffic of the spinal cord**. Interestingly, the eyes and the spinal cord are a direct extension of the human brain appearing early in embryonic development, whereas the protective conduits and grooves provided by the hard skeletal system are developed later **during the fetal stage**, all accurately mapped for the complete network of nerve fibers and blood vessels!

4.39 The level of complexity these examples purport of the principles (so far) of Dynamic Evolution makes the illustration of the Eccentricity Diagram (discussed in Part III) **look like a children's math lesson in elementary school** due to the dependency cascade.

4.40 Therefore, the lattice-map of this tiny section of the Plexus has **multiple connections** for these (and other) examples and therefore **multiple dependencies** that span (and sometimes *jump*) the levels of the Plexus.

4.41 So now our above city-sized analogy takes on another new dimension. These stones do not have *meaningless* words inscribed on them. Parad-oxically, each stone uses an elegant economy of words that provide **instructions on how and where to lay down that very stone**!

4.42 If our intrepid travelers on this untouched island found the question of "How can we explain the engraving on this stone?" to be compelling, what would they say if they encountered this city-sized collection of stones? The complexity level has risen a little more...

4.43   In the spacecraft analogy above, the comparison would go like this: each component manufactured by each employee at all levels, when completed, automatically bears the instructions for the assembly of another component identical to itself! When asked 'How do you make this component?' **the employee simply points to a completed component and reads out the instructions printed on it during its manufacture!**

4.44   This chicken and egg question is a thorny problem for Darwinian Evolutionists who try to explain, for example, how RNA (which carry instructional messages from DNA) made protein without having the help of any pre-existing proteins, and how proteins made RNA without pre-existing RNA. However, within the context of the Plexus, the *multiplicity* of key dependencies for RNA and DNA and its associated specific proteins stretch the "It-came-from-something-simpler" explanation to an absurdity analogous to the *width* of a literal single strand of DNA itself compared to the *length* of all the DNA strands in all the human bodies on earth! A further analogy would be with our Space Shuttle (§4.11). The components that relate to one another, in terms of proximity, multi-faceted similarity, or functionality, are mapped by sub-parts, paint color, form and shape, size, contextual purpose and fit, and a variety of other criteria written down in the detailed specifications. DNA contains all of this information and a great deal more.

COHERENCE
THROUGH
CONNECTIVES

5   COHERENCE THROUGH CONNECTIVES

Outline

5.1   So far we have made repeated references to the various *connection types* in the Plexus (see §3.18 for an overview). The logical arrangement of the Entities, and the connections between them, is one of the fascinating aspects of Dynamic Evolution. We will outline just a few of these in this chapter.

Configurative Connections — Fractal Patterns

5.2   The repeating patterns that occur throughout the Plexus are an intriguing aspect of the Dynamic model. Whatever size we consider, and no matter what time-scales, there are similarities of form, nature, or function that are particularly striking, aside from their application to the advanced features of Dynamic Evolution.

5.3   At this point, a comparison with the claims of some advocates of Darwinian Evolution serves to demonstrate the superiority of Dynamic principles.

5.4   For example, the gradual development of the human embryo into a full-grown baby is often used as a *parallel* of nature's phased development of human evolution by natural selection. As the comparisons in this section demonstrate, there are too many difficulties with this explanation when **compared with the contextual fit and the compelling argumentation of the Dynamic model.**

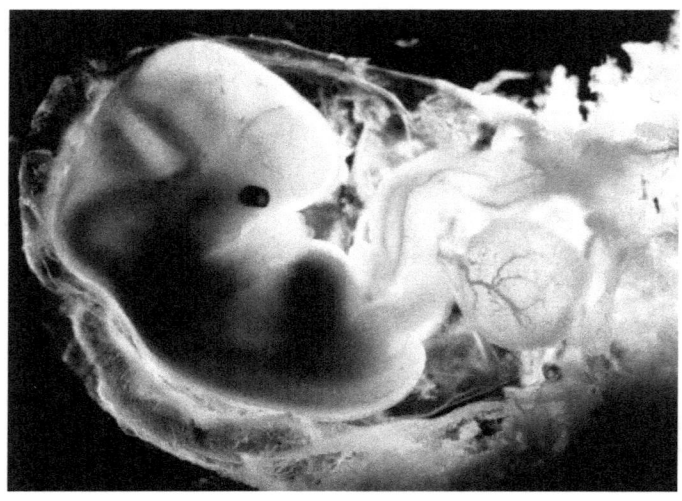

Figure 20 - Human embryo at 6 weeks

Figure 21 - Illustrating computer software design (picture courtesy of cavsi.com)

5.5 One difficulty becomes clear with the following consideration: In the making and design of the Nasa Space Shuttle, discussed in the previous chapter, the various preparatory models the designers used for testing different parts during the design and test phases served as

patterns for the completed product. In a similar way, when a tailor designs a suit, the pattern for the finished product is mapped out on the paper he or she uses (a template, or tracing paper) prior to the start of the manufacturing stages. In the design of a computer program, the programmer creates a number of small modules, each performing a minor but necessary distinct task; other larger modules make use of these to perform major tasks; this progressive development continues until the "parent" program makes use of all the key modules as it provides the overall functionality for the application, or *app* as they are called these days.

5.6 Going back to our discussion of the development of human life within the womb: The stages of human development from conception to birth are like the *patterns* discussed above. Neither the embryo nor the fetus nor the newborn baby, is a "finished product." They are **immature versions of the adults they are destined to become**. The complex patterns that "mapped out" the complete set of features, attributes, and parameters that define this person as an adult, were written out in another form long before birth.

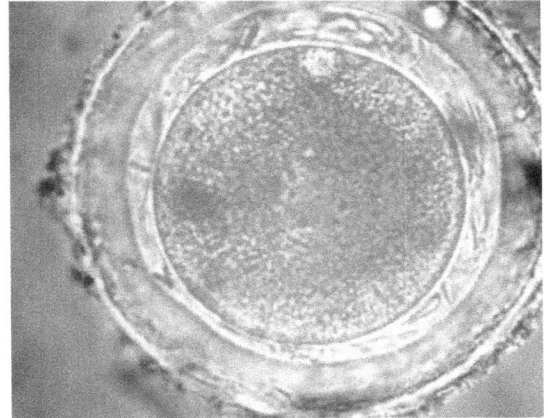

5.7 The DNA *pattern* that gave every developing cell the required information to grow, reproduce, and function, and ultimately to build the complete body, albeit very tiny initially, was completed only minutes after conception!

5.8 These "patterns" are discerned (consciously or subconsciously) and appreciated by the child's parents.

They see in the baby-features of their child a kind of fore gleam, or fore-view, of the child's intended appearance. This "fore-view" may not be apparent until the child reaches adulthood and the parents reminisce on old photographs. It then becomes clear that those features that made the child so endearing, so cute, at all phases of growth and early development, were **immature versions of the adult features they now possess**. How often have we seen a parent gazing appreciatively and lovingly at their young offspring, wondering what he or she will be like when fully grown? Yet the characteristics possessed by the adult human were written out in the zygote (see §5.11)!

5.9 The above principles are manifest in almost every creature on earth, indeed in the **development of all life**. However, the principles fit snuggly into the explanation of the development and *growth* of existing life, but too many problems occur when trying to make these same principles fit the *origin* of life using Darwinian principles.

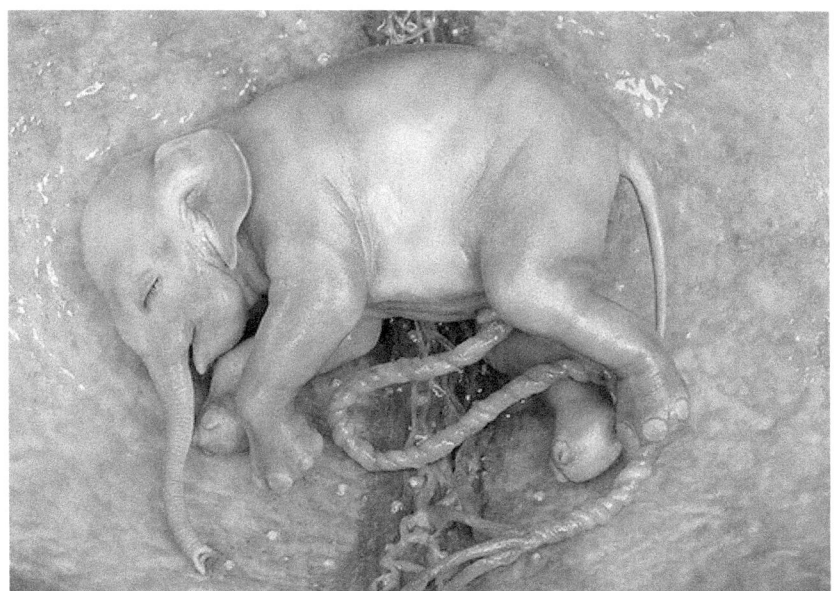

Figure 22 - A baby elephant at fetal stage (photo by anroir)

5.10   This tiny elephant would probably not be able to survive at this stage outside of its mother's womb. A great deal of further development must take place before it can emerge successfully into daylight. However, even after birth there is usually much to learn for any species, as the next photo demonstrates...

Figure 23 - Even for mighty young elephants, there is much to learn
(Photo courtesy of www.saveelephant.org)

5.11   This serves to underscore the failings of the Darwinian argumentation. Unless the elephant survives every phase of its long development (up to two years in its mother's womb), it will not be around to perpetuate the species! The early development phases are a means to an end:

- The zygote (1st cell developed from conception) contains a "written" pattern for the finished product as well as a set of instructions for embryo development
- An embryo is a pattern for a fetus
- A fetus is a pattern for a newborn baby
- A baby is a pattern for a young adult
- A young adult is a pattern for a mature adult
- A mature adult is the finished product

Configurative Patterns of the Plexus

5.12 This concept of patterns, or configurations, that are evident in the multiple stages of a developing creature in its mother's womb, bear a striking similarity to the patterns of the Plexus, manifest throughout the stages of the development of the universe as well as through all the 12 levels of the Plexus from the tiny quantum particles to the largest galactic superclusters. These patterns, or configurations, often occur in a fractal-like arrangement, as we have seen.

5.13 Let's take an example using progressive stages in terms of size:

- The particle patterns formed at the sub-atomic level
- The patterns formed by electrons around an atomic nucleus and by gluons within the nucleus
- The levels (steps) that confine energy output and input to particular stages, for example electron shells/sub-shells and heat exchange
- The bonding patterns formed by atoms in the construction of a stable molecule
- The patterns of proteins which are organized by a fixed menu of folds determined by geometry
- The bonding patterns formed by molecules when they are assembled into a living cell
- The patterns formed by cells as they combine into a complete organ
- The bonding patterns formed by organs, organized into a complete body

5.14 But these configurations bear a similarity of form and function in other ways:

- The gluon-cloud patterns in the atomic nucleus and the atom itself with its array of electrons in fixed patterns around it

- The molecule with its bonded-atoms and the invisible electromagnetic activity around it

- Planet earth with its internal activity and outer protective magnetic shield

- The earth-moon system and the mutual gravitational attraction

- The solar system with the sun at its center and the planets, asteroids, and other objects bound to the sun by gravitational forces

- The galaxy with a black hole at its center and the body of stars, stellar material, gases, and other objects revolving around it

- The Galactic Wall with the symphony of galaxies revolving in elegant formation

- The supercluster of Galaxy clusters revolving around a gravitational fulcrum

Note the illustrations on the following pages, demonstrating the progressive examples of a tiny sample of the configurative patterns of the Plexus.

Figure 24 - Pattern 1: Paths traced by quantum particle collisions

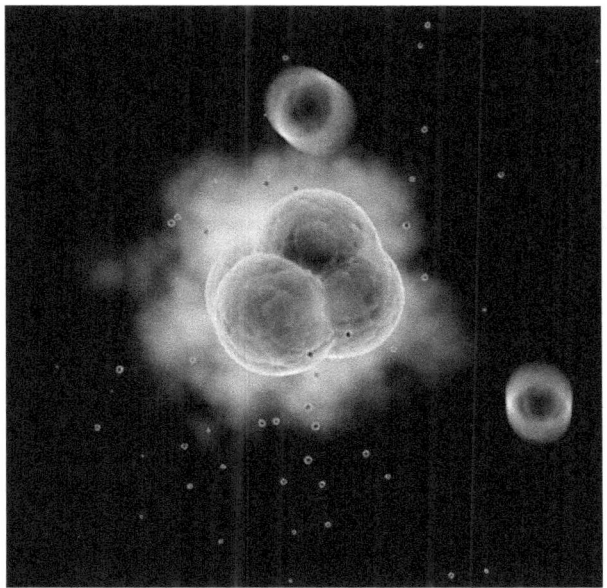

Figure 25 - Pattern 2: An atomic nucleus with "orbiting" electrons

Figure 26 - Pattern 3: A portion of a molecule of ethylene (courtesy of Nasa)

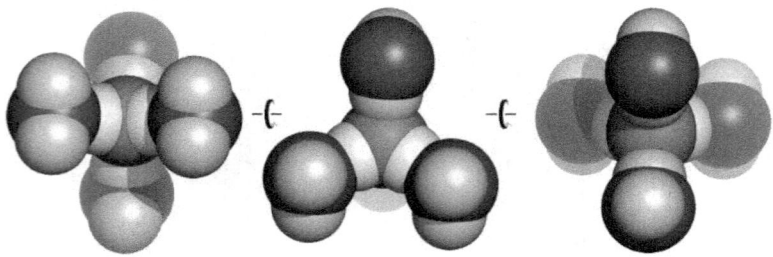

Figure 27 - Pattern 4: Five water molecules in a hydrogen-bond (two bonds are donated) (Picture courtesy gatech.edu)

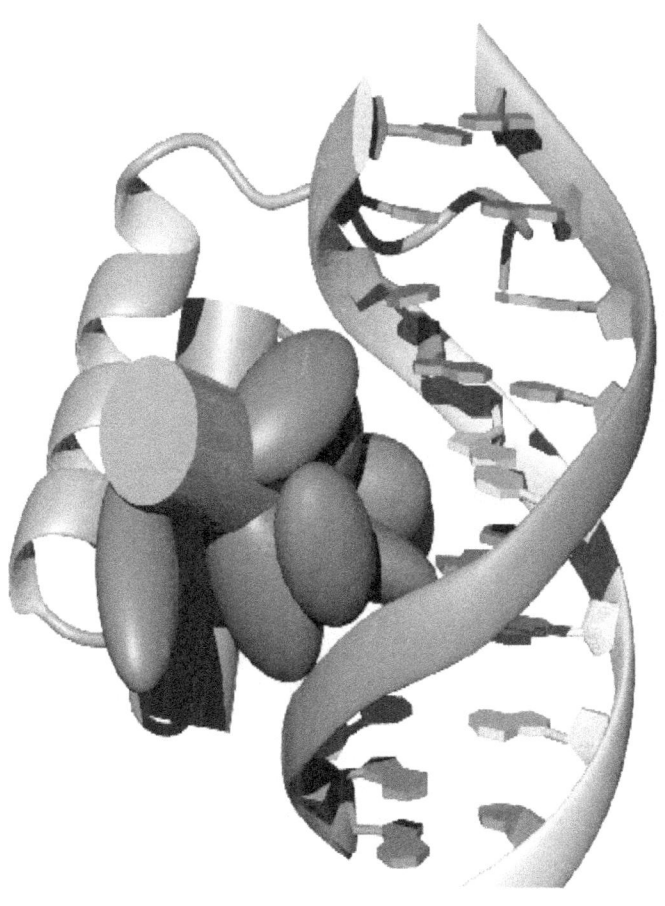

Figure 28 - Pattern 5: DNA, a molecule made from molecules
(Picture by "The MolMol")

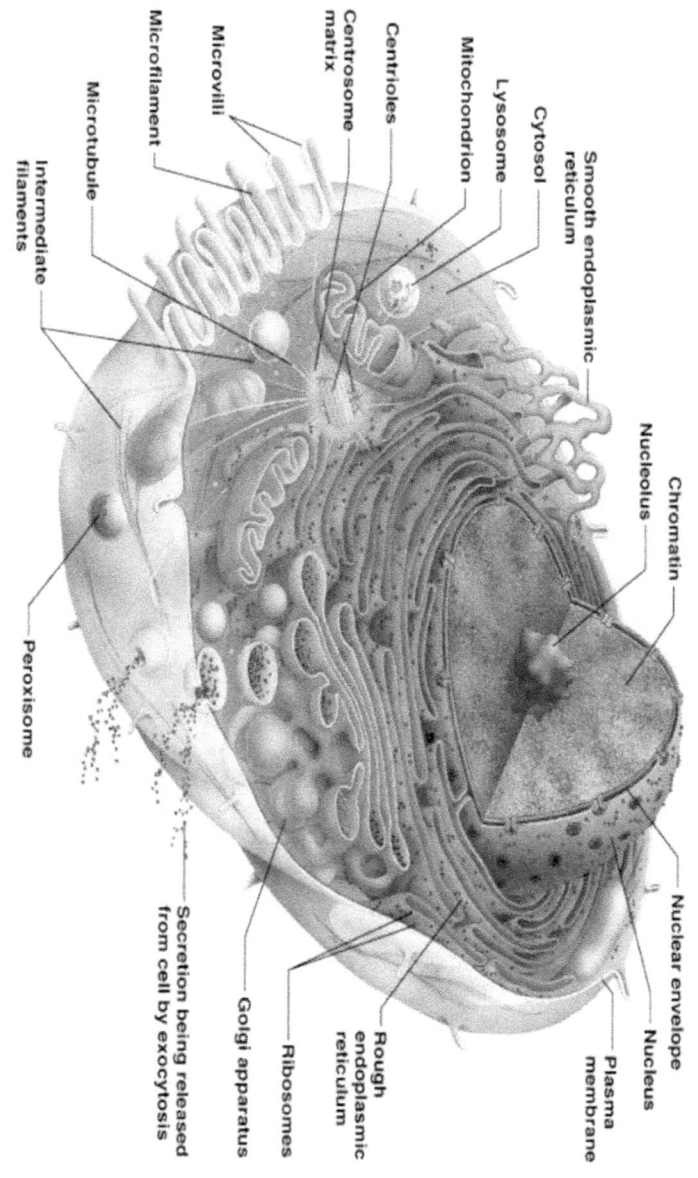

Figure 29 - Pattern 6: The cell, containing numerous molecular machines

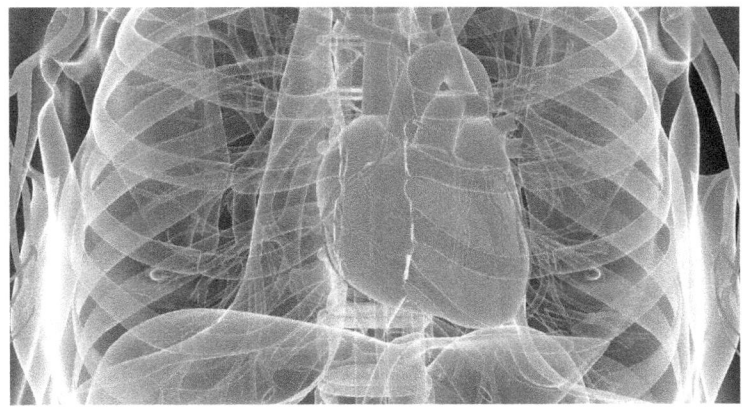

Figure 30 - Pattern 7: The heart, centric to the body's organs

Figure 31 - Pattern 8: Gravicentric earth-moon system (courtesy of psi.edu)

Figure 32 - Pattern 9: Our sun-centered system of connected objects (Nasa.gov)

Figure 33 - Pattern 10: The gravicentric Milky Way
(outerspaceuniverse.org)

Figure 34 - Pattern 11: Galaxy Cluster Abell 2667

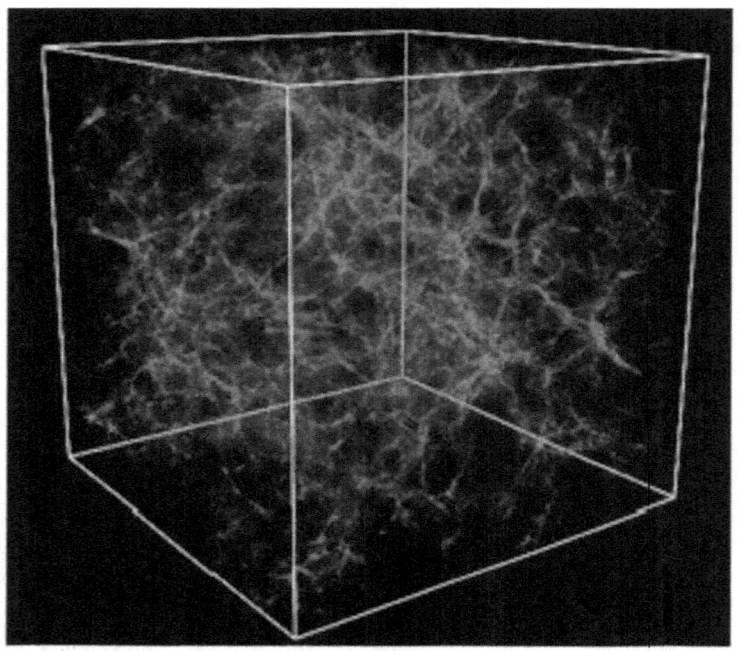

Figure 35 - Pattern 12: Galactic Superclusters

5.15   The above 12 illustrations roughly correspond to the 12 levels of the Plexus. There are a number of **key characteristics that each of these items share**. For the most part, there are three distinct features that can be stated thus: (1) a center that provides either a control function or a centralized influence over (2) the surrounding area which contains functionality intrinsic to the item, and finally (3) an outer protective layer or mechanism. These characteristics, or physiognomies, are summed up in the following table:

Fractal Configurations of the Plexus — 1

ENTITY	CENTER	MAIN BODY	OUTER SHELL
Atom	Nucleus	Identifying orbiting electrons	Concentric shell layer (electron orbitals) & magnetic field
Molecule	Central covalent atomic nuclei	Shared orbiting electrons	Affective and effective electromagnetic interactivity
Cell	Controlling nucleus	Multi-functional cellular machines	Protective membrane
(Some) Organs	Highly-functional defining central utilities	Complementary functional components	Protective outer layer
Head	Brain	Functional and protective components	Protective outer layer, the skull
Torso	Principle body organs	Complementary functional organs	Protective outer shell—rib cage & spinal column
Human Body	Brain	Complementary functional organs	Protective outer layer—skull, flesh, and skin
Earth	Functional central core, magnetic field generation	Functional body: heat dispersion, material recycling	Protective outer shell, earth's crust, atmosphere, and magnetic field
Solar System	Dominating centric star, the Sun	Sun-centric arrangement of solar bodies	Protective outer magnetic field
Galaxy	Dominating centric black hole	Gravitationally powered star systems	Protective outer magnetic bubble
Super / Cluster	Gravity fulcrum	Gravitationally powered galaxies	Protective outer magnetic bubble

Configurative Connections — Functional Patterns

5.16   We have seen in the above sub-section that a striking feature of the Plexus is the fractal-like appearance of its components throughout all of its 12 levels. Later studies will expand on this topic: for example, the discussion of aquaporins in §4.3 identifies the complex functionality of these highly specialized proteins, but the truly fascinating aspects of this study involve the **striking comparison of their counterparts in a variety of living creatures**. Advocates of Darwin are hard-pressed to explain the presence of these same proteins in organisms as diverse as animals and plants. **Did Darwinian principles select the various components of these proteins over and over again** for all the samples found in nature? This is one of the Dynamic principles that casts large shadows over the Darwinian model.

5.17   These functional patterns are not necessarily of the visible kind, as in the previous sub-section, but are closer to the analogy of specific machines. A simple example, one that 'jumps the levels of the Plexus,' (see §4.40) is the functionality of the **strong nuclear force** within the nucleus of the atom, and its counterpart at the other end of the size scale, **gravity**, which likewise produces an attraction, albeit on a very considerable scale of size with a corresponding decrease in power in comparison. A slightly more advanced example (for the physics student) is a **comparison of the particles employed by three of the fundamental forces**: the (1) strong nuclear force, (2) electromagnetism, and (3) gravity; namely (1) gluons, (2) photons, and (3) gravitons respectively.

Tubes — Not Strings!

5.18   Many other examples are given using features of the human body, which makes extensive use of shapes, for example *tubes*. You may remember that we referred to tubes in our discussion of the Space Shuttle (section 4.13) and the conduits that transport water, power, waste, and other items, through the ship. Tubes are found within the cell (e.g. aquaporins, microvilli, centrioles, microtubules, and some mitochondria), within organs such as the lungs, heart, kidneys, and the circulatory system, within the central nervous system, the muscular system, the digestive system; but they are also found in the plant kingdom, for example **pollen tubes** that transport male reproductive cells through female flower tissues for fertilization, **sieve** tubes that transport vital carbohydrates, and **tracheid** cells and **aquaporins** which transport water.

5.19　The heart itself contains more "tubes" than are generally known by the layman: it contains **a network of capillaries to enable its own tissues to be fed with blood**. This begs the question of the Darwinian advocate: Which evolved first, blood vessels, blood itself, or the central organ of the cardiovascular system, the heart? The explanation of Dynamic principles applied to the development of the heart (which is the first organ to develop in the embryo) is the superior explanation. For if we map the development stages of the human body to the Darwinian model, how could we explain the gradual development of an organ that did not yet possess the fluid that it was designed for?

5.20　The aspect of "disparate functionality" is often unappreciated by Darwin adherents. **Survival of the fittest provides an exceedingly narrow margin for the scientific explanation of so many disparate yet interdependent functions**. A considerable number of the functions provided by the parts and organs of the human body have little to do with survival. For example, why do animals have bladders or rectums? If we did not happen to have a bladder, we would simply expel a "squirt" of urine rhythmically every few seconds. Would we still survive? It might be somewhat annoying, but **how would**

Darwinian principles come into play? Would there be some mechanism that "ordered" the development of a bladder? If so, to what purpose? It may best be described as a social feature; possessing a bladder, or a rectum, enhances the quality of life by a small amount, enabling us to enjoy life much of the time and defecate infrequently. If we were scrambling for our very existence, determined to be the creature that survives all the rest, would we care that we defecate so often? And if we did, how would nature accommodate our wish?

5.21 Similar questions could be asked of the stomach. Without this organ, we would need to eat food every half an hour or so, but we could survive! Possessing a stomach enables us to enjoy hearty meals, but then to get on with our lives whilst the digestive process goes on without any further conscious effort on our part.

5.22 The tube-like formation of these body parts and their associated appendages, and the parallelism manifest in the functionality of other tube-like organs and parts, is **a feature intrinsic to Dynamic Evolution.**

5.23 An instructive example of tubes occurring in other areas of nature occurs on the Big Island of Hawaii in the *Thurston Lava Tube*.

Figure 36 - The Thurston Lava Tube (Nahuku) at Hawaii Volcanoes National Park

5.24   The Thurston lava tube was formed gradually over a period of many years by the flow of low viscosity lava producing a continuously hardening crust around the edges of the tunnel. Successive lava flows added to the crust, with the final result giving the appearance of a man-made tunnel. The interesting nature of this tunnel, its development by the force of a flowing molten liquid, provides a **comparison with the argumentation of Darwinian principles in opposition to Dynamic principles**. Taking a step back for a moment: Which would Darwinian exponents say evolved first, blood vessels, or blood? We have discussed the impressive paths taken by developing blood vessels (and their relation to developing nerve fibers) in §4.33, now we see this argument through another facet (the linkages of the filaments of the Plexus). **How does the Darwinian adherent explain the gradual development** of (a) the heart, (b) the blood vessels, (c) the lungs, and (d) the blood itself? The natural path of the molten lava, cutting a tube-like formation in the rock, might be a "natural" place to look for an explanation (the forceful path of blood flow), were it not for the obvious problems either a lack of blood, or an uncontrolled blood-flow, would cause! Are we looking at the independent development of almost 100,000 miles of blood vessels, in parallel with both the heart and the lungs, as well as the complex consistency of blood itself? Did these organs function adequately prior to having their required blood supply? Did the heart form an attachment to the lungs prior to the development of blood, or subsequent to? How did this tube-like attachment (the pulmonary artery) come about? Were there many tube-shaped protuberances available from which Darwinian principles selected the best fit? Or is it not rather a demonstration of Dynamic principles?

5.25 For this system to work at all, there needs to be a rich blood supply, along with an apparatus for oxygenating the blood. A mechanism for producing blood and its components needs to be in place, as well as the necessary conduits to transport the fluid to the target sites. And of course all of this must be considered in concert with the highly complex constituency of blood itself. What is necessary is a complex set of principles that are far superior to the concept of survival of the fittest.

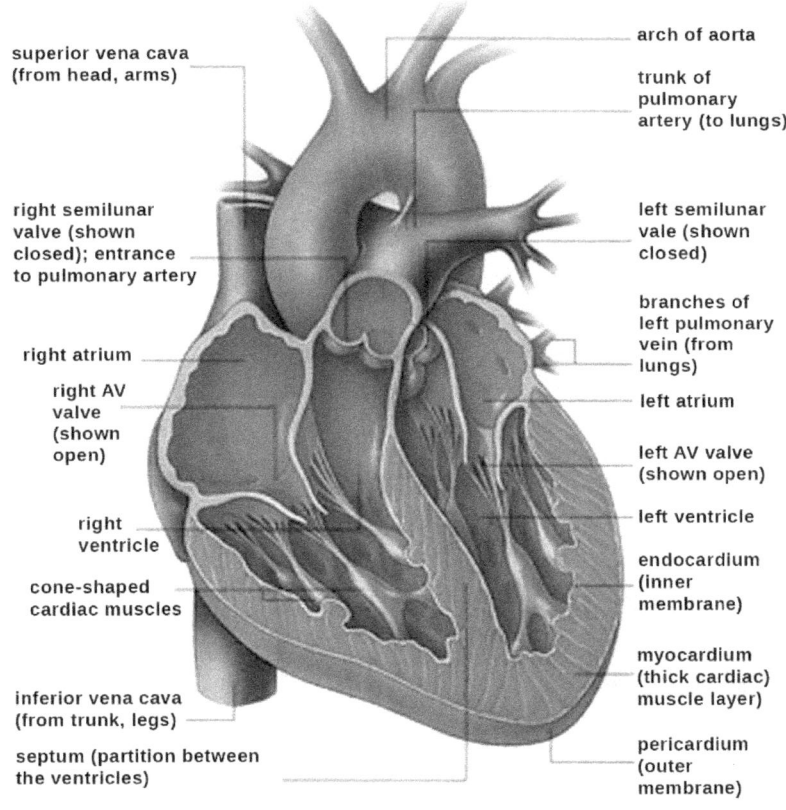

Figure 37 - The human heart, showing some of its inter-connected tubes

Muscle Fibers — Tubes Within Tubes

5.26   The extensive system of muscle fibers throughout the human body provides an **example of the Dynamic principles intrinsic to the tubular shape**. There are approximately 640 muscles throughout the body and almost all are symmetrically distributed between the left and right sides. Like the Russian doll with multiple dolls embedded, one within the other, muscle fibers find their versatility and strength in the use of *tubes*.

Figure 38 - "Fractal" Russian Dolls, the Matryoshka

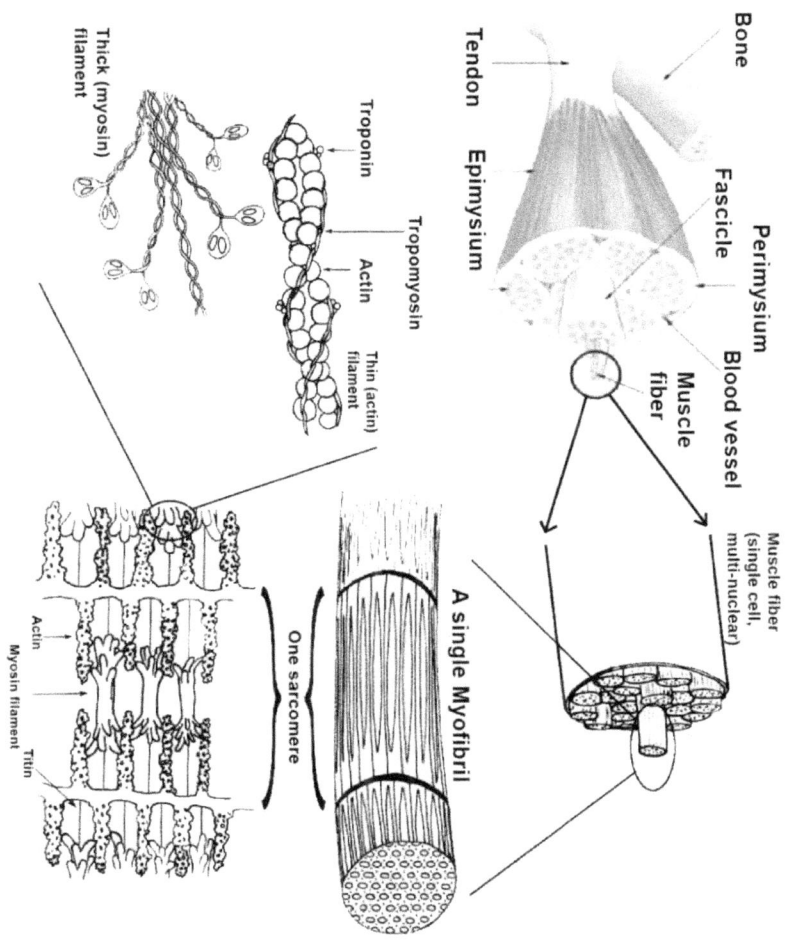

Figure 39 – Similar to the design of the Russian Doll, muscle fibers are composed of tubes within tubes

5.27 The illustration above provides a simplified view of the structure of muscle, down to the microscopic molecular myosin proteins, which themselves provide further examples of multi-function Entities (§4.18) — more than

17 classes of myosin proteins have been identified to date. These motor proteins are involved in muscular contraction, cytokinesis (cell division), short-range membrane/vesicle transport, and many other cellular processes.

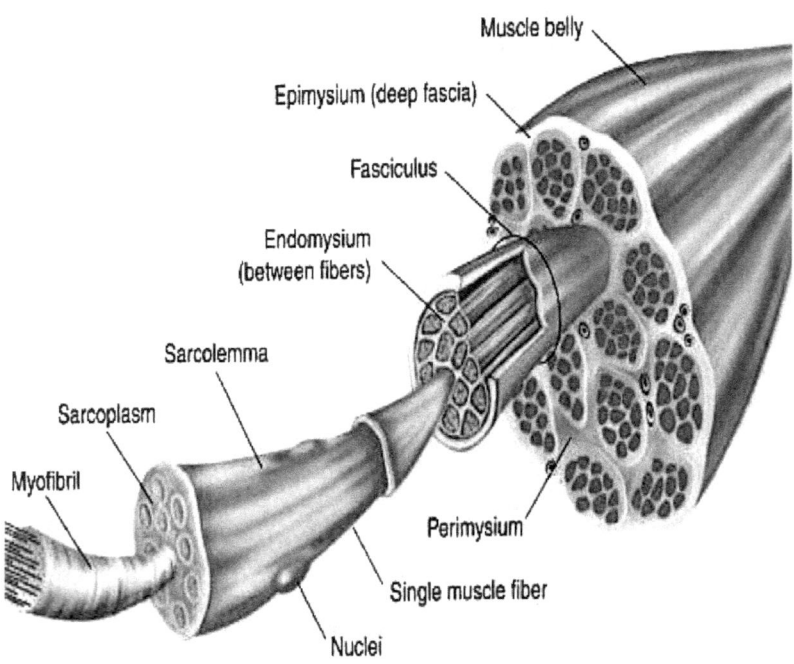

5.28   The telescopic formation of the components of muscles may be summed up in the following list:

(1) Enclosing epimysium (attached to tendon)
(2) Fascicle
(3) Muscle fiber
(4) Myofibril
(5) Myofilaments

5.29   This list, though, is in descending order of size. During the manufacture of the Russian Dolls, illustrated above, which would you suppose would logically be made first, the largest Doll or the smallest? Likewise, during the development of muscle fiber in the human embryo, which would you say is developed first, the smallest components of myofilaments (the actin and myosin) or the largest? And which might the Darwinian proponents say evolved first?

5.30   In the case of the Russian nested Dolls, called Matryoshka, the smallest Dolls are apparently manufactured first. All the Dolls are fashioned from the same cut of wood, to minimize any shrinking/expanding difficulties later. The master Matryoshka maker carves the Dolls on a lathe, but does not use any measuring instruments; he uses his expert judgment to match up each one. In the completed set, the smaller Dolls each form a perfect fit into the larger Doll next in size.

5.31   This seems to be a logical way to develop such a system; ensure that the smaller component is made first, then progress to the largest. But **this is not the way muscles are formed within the developing embryo**. The harder protective layers, the sheathing, the epimysium, perimysium, and endomysium, develop first, and afterwards the cellular filaments form. This is reminiscent of the **inside-out development of some brain cells** discussed in §4.22.

5.32   Once again, the evidence supports the principles of Dynamic Evolution; **Darwinian Evolution appears to be a poor fit**.

5.33   There are far too many additional examples, within the human body alone, to include in this short work, which covers just the fundamentals of Dynamic Evolution. Further examples in later studies include: the semicircular canals and the **cochlea** of the ear, the efficient **shaping of bone**, tubular cells, the use of tubes throughout the digestive system, the lungs, the mammary glands, the reproductive system, the salivary gland, the pancreas. Also, the use of the tube shape in other areas of nature is examined: the chaetopterid worm, polychaetes, the sea angel *Clione limacine*, and insects like the Cinipid Wasp:

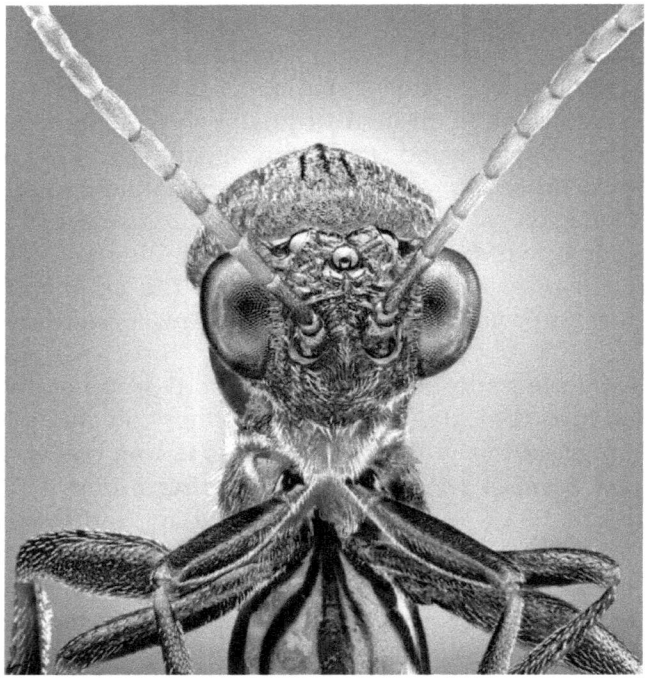

Figure 40 - The versatile, multi-eyed, Cinipid Wasp, pictured by Tomas Rak, courtesy of focusingonwildlife.com

5.34   Additionally, the tubular systems in the common fly are compared to those within the human body; and the remarkable similarities of design are highlighted. No-one would suggest that humans are in some way descendants of the common house fly. But the multitude of design similarities reinforce the position of Dynamic Evolution as the superior explanation to the origin of life.

5.35 The *Dynamic* features of the *tube* shape are manifold. Later studies enlarge on the use of this shape and include other geometric consistencies in nature. Time and limited space do not permit an extensive treatment here; but you are encouraged to seek out the physicists' discussion of toroidal geometry and its place in the "schema," or what we would here refer to as The Plexus.

Machine Codes
of the
Plexus

6   MACHINE CODES OF THE PLEXUS

Outline

6.1   Many comparisons have been made between aspects of nature and computer science. Even now there are scientists working on the use of the principles of quantum mechanics to build faster and better **computers**. Other work includes the use of DNA in the storage of **vast quantities of information**. Still other scientists are using what is known of the human brain and its cognitive abilities and the neuronal basis of higher functions of the brain in order to build computers of tomorrow, including machines with AI (**Artificial Intelligence**).

Figure 41 - A scene from 'The Matrix' - A film by The Wachowski Brothers

6.2   The enormous storage capacity of DNA is now widely known. In a recent experiment, researchers stored an entire science textbook in less than a *picogram* (one trillionth of a gram) of DNA (*New Scientist*, August 2012). This means that **one gram of DNA could hold 700 terabytes of data** (new PC computers today tend to come with a single 1 TB disc installed), or over 13,000 Blue-ray discs. The computer company Google recently *counted*, using its own specific algorithm, the number of book "titles" in the world at the present time. The number quoted was just short of 130 million. Yet, according to the above experiment with DNA, a single gram could contain the text of *one trillion*—one million times one million—books.

The Languages of The Plexus

6.3   One area of computing technology that finds many parallels within The Plexus is that of "machine code," the "language" of computers. Computers are often referred to as "stupid" machines, even by their inventors. This is because they process enormous amounts of tiny pieces of information in order to achieve any kind of meaningful complex result, and they must be told, "programmed," every step of the way, how to perform each and every task. And, as any programmer will tell you, it takes less time to write the initial "code" for a program than it does to find the many "bugs" that stop the program from working efficiently, or at all.

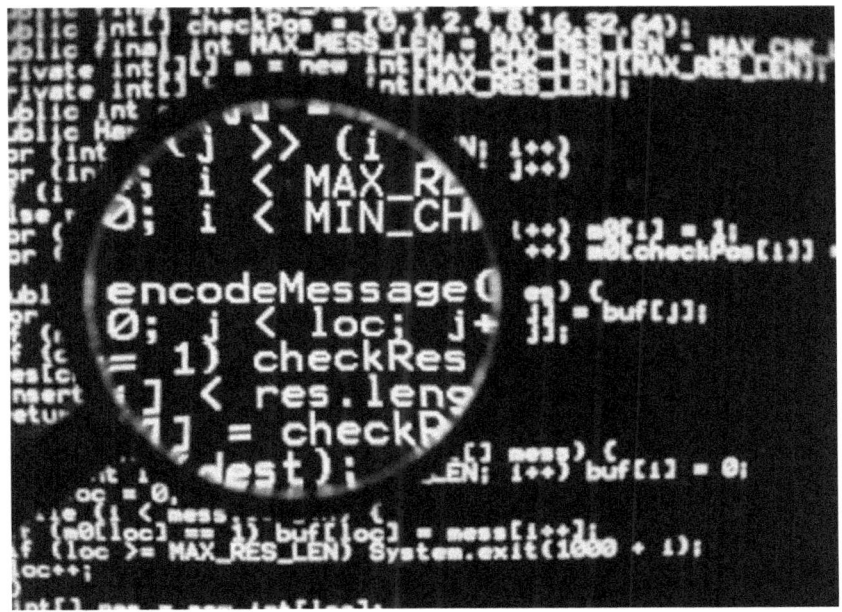

Figure 42 - Machine Code (a "lower level" than the above code illustration) may be unintelligible to most humans, but is required by most computers today

6.4　　Programmers are trained communicators, experts in translating human requirements into "code" that a computer can act upon. The computer code, or "language," a programmer uses, has a basic syntax and morphology that uses progressive, logical steps to achieve any desired result. One such method is the use of "variables" to calculate values. The programmer can invent a name for a variable and then assign a numeric value to it. Using this technique, a variety of operations can be performed. The programmer might, for example, assign a variable called BLUE a value of 23, and a variable called RED a value of 41. In the code, the line:

　　　　BLUE+RED

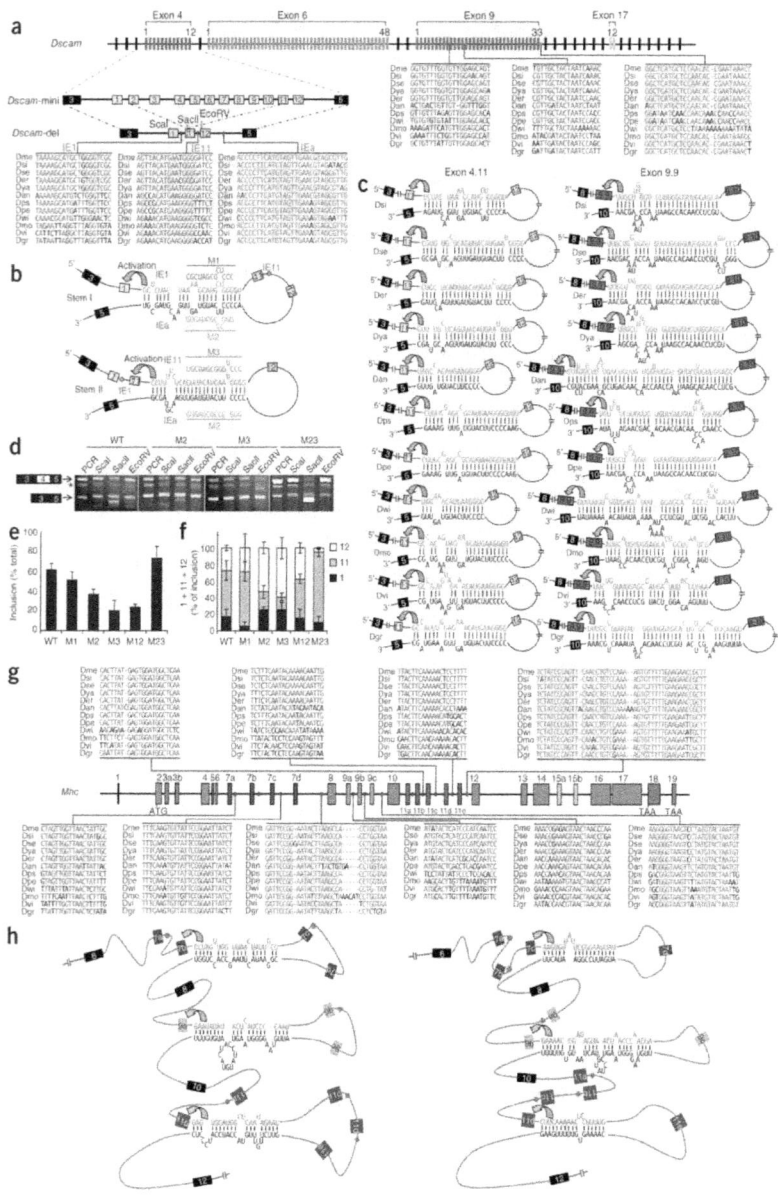

Figure 43 - DNA, the "machine code" of life

would yield the value 64. If the values of either BLUE or RED change, then the result will change accordingly. This simple arrangement can be carried through to more complex formulas that calculate the value of a shopping list, for example, or the percentage depreciation of a company asset.

6.5 The programmer uses a "code" to build the required formula and to get the computer to either display results onto the monitor, print them onto paper, or get the computer to save the results for later use. All of these operations find their counterparts within The Plexus. **Machine code is found in the behavior of quantum particles**, including the orbitals and energy levels of electrons and the interaction of quarks; in the ordered structure of the periodic **table of elements**; in the behavior and interaction between molecules, particularly **proteins and enzymes**; in the "language" of **DNA**; in the signaling of the electro-chemical and quantum **probabilistic pathways of the human brain**; in the mathematical precision of the laws that govern planetary, stellar, and galactic objects. These are expanded in later studies, for example in the full chapter *Machines of the Plexus*.

Quantum Language Code

6.6 The nature of some of the smallest "machines" within the Plexus can be demonstrated from a further study of the atom. Nuclear fusion is the process that builds heavier atomic nuclei from lighter ones. The nucleus of the atom consists of two main components, protons and neutrons (which in turn consist of quarks). Just as the atomic nucleus is surrounded by a "cloud" (that is, the "orbiting" electron/s), so too quarks are surrounded by a cloud of gluons (which are the exchange particles for the "color" force between quarks; this force has a logical Plexus link to the exchange of photons in the electromagnetic force between two charged particles).

6.7   Protons and neutrons make up the nucleus, held together by the strong nuclear force (which acts upon quarks), which even overcomes the powerful repulsive force that should keep protons apart. The proton is a baryon and is considered to be composed of two "up quarks" and one "down quark." Baryons are made up of three quarks in the standard model. This class of particles includes the proton and the neutron.

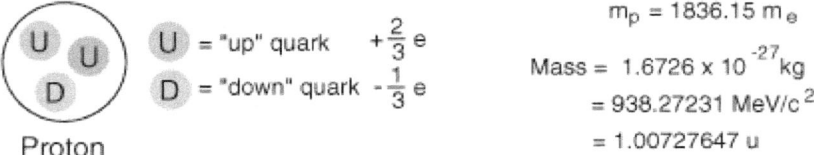

$$m_p = 1836.15\, m_e$$
$$\text{Mass} = 1.6726 \times 10^{-27}\, \text{kg}$$
$$= 938.27231\, \text{MeV}/c^2$$
$$= 1.00727647\, u$$

U = "up" quark $+\frac{2}{3}$ e
D = "down" quark $-\frac{1}{3}$ e

Proton

6.8   The above is just a fraction of the study of the world of the sub-atomic particle. However, some of the *machines of the Plexus* have already been touched on. Note for example the behavior of the strong nuclear force in binding the quarks that comprise the particles of the nucleus together. This forms a "connectivity of function" with the force of gravity, which acts on objects many times greater in size. It also has a "connectivity of form" correspondence with the electromagnetic force. Just as magnetic poles of opposite polarity attract each other, and the invisible force of electromagnetism acts upon both objects, so too the invisible strong nuclear force ensures that the protons and neutrons at the heart of the atom form an unbreakable bond.

Periodic Table of Elements

*Lanthanide series
**Actinide series

Language Code of the Elements

6.9 As any student of chemistry knows, all known material things are composed of various combinations of about 100 different elements.

6.10 In 1913 British physicist Henry Moseley confirmed that an element's chemical properties are only approximately related to its atomic weight (roughly equal to the number of protons plus neutrons in the nucleus). What really matters is the element's atomic number—the number of protons its atom carries. Since this time, elements have been arranged on the periodic table according to their atomic numbers. The structure of the table reflects the arrangement of the electrons in each type of atom.

6.11 Electrons may only orbit a nucleus using specific "orbitals" or shells. These are referred to as s-orbital, p-orbital, d-orbital, and f-orbital. Electrons exist on these specific shells, and this is what determines their energy level. The Periodic Table shows the logical arrangement of elements based on the number of protons in their nucleus, the number of electrons, and the orbitals.

6.12 Most elements in the Periodic Table are metals. When metallic atoms bond together, the electrons that exist in the outermost orbitals of these elements become free, or *delocalized*, and are shared between the nuclei of all the metal atoms. Now, instead of only being allowed specific energies, they have a continuous range of possible energies, *conduction bands*. This free movement of the electrons ensures that metals are generally good conductors of electricity. You can find the metals neatly grouped on the left-side and the center of the Periodic Table. Atoms that do not share their electrons freely, for example in the element *sulfur*, are good insulators.

6.13	The properties of the elements are far too numerous to list in one volume. But the material covered so far demonstrates the fixed attributes of the elements, and the simplicity and elegance of the arrangements that give them their particular properties. If sodium is placed beneath lithium and not next door to fluorine, and potassium is placed beneath sodium to begin another row, and so on, the vertical lines of elements are chemically similar. These vertical lines are called *groups*.

6.14	The logical order manifest in the table of elements is due to the consistency of Dynamic principles. The Periodic Table is sometimes referred to as the Periodic Law by scientists. The principles that a study of this table teaches, provide the same consistency and regularity of pattern that other aspects of The Plexus demonstrate, as later studies will show.

Language Code of DNA

6.15	The basic units that make up DNA are called nucleotides. Each nucleotide contains one of four chemical bases: (A) adenine, (C) cytosine, (G) guanine, and (T) thymine. DNA's code is written in only these four 'letters', A, C, T and G. The meaning of the code lies in the sequence of the letters in the same way that the meaning of a word lies in the sequence of alphabet letters, then a collection of words produces a sentence, and a set of sentences provides a comprehensive, intelligible, set of instructions. Your cells read the DNA sequence to make chemicals that your body needs to survive.

6.16	A gene is a length of DNA that contains the instructions to make a particular chemical required by the body. The DNA in a gene usually codes for a protein.

6.17	Proteins are the building blocks for most of the parts of the human body.

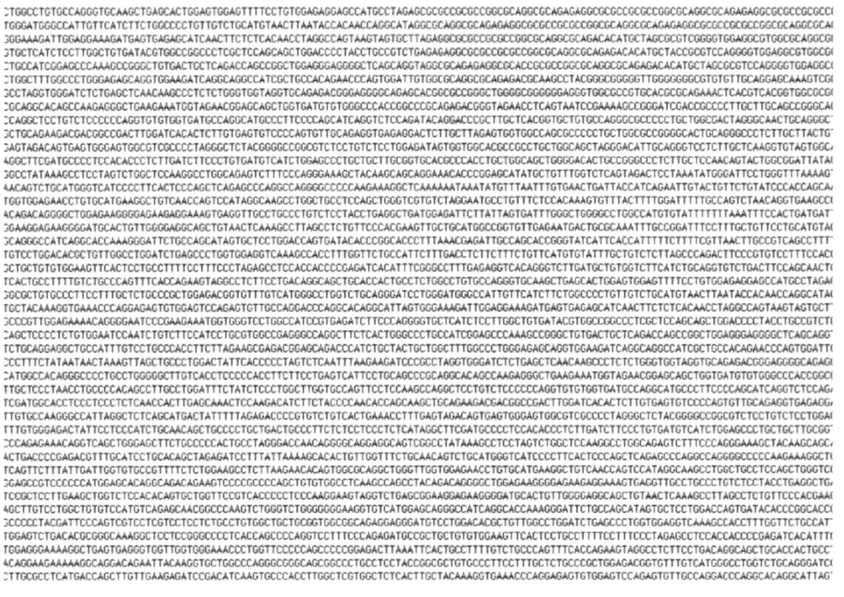

Figure 44 - A section of specific DNA code

6.18   The entire package of information stored in the DNA is called the *genome*. It is about three billion "letters," or nucleotides, long. If it were transcribed onto paper, the book would fill some 200 volumes each the size of a 1,000-page book, according to the Human Genome Project.

6.19   In more advanced studies in this series, we will deal in greater depth with the languages of The Plexus, for example: the Language that governs the behavior of all universal particles large or small; the Language of neuronal message transmission in the human brain; the Language of human communication; the Language of planetary and stellar interaction and other astronomical laws. For example, one fascinating discussion centers on the discovery by neuroscientists of two separate areas of the brain that are **hard-coded to deal with human**

language. One area handles the complexities of incoming human speech, and the other deals with the construction of human speech to enable us to build intelligent sentences; otherwise two-way communication would be extremely difficult.

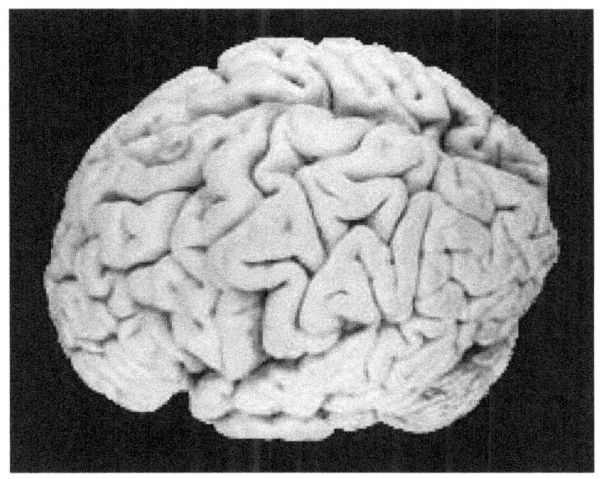

The Mastermind Effect

7 THE MASTERMIND EFFECT

Figure 45 - The Mastermind Game by Invicta Plastics

7.1 Darwin's theories, associated with the principles of Natural Selection, can be likened to players in a game of Mastermind (a game of logic by Invicta Plastics Limited). You may have come across this game which includes colored counters (or pegs) that one player arranges sequentially in a code. His opponent has to crack the code, first by guessing the choice of counters, then, on subsequent attempts, by following the clues the code-maker provides. The true code, set by the code-maker, we can liken to scientific truth, i.e. the way things really are according to generally acceptable "norms"; or facts, if you prefer.

7.2 The attempts to break the code, we can liken to the theories associated with Darwin's survival of the fittest. It is not incorrect to say that these theories are "attempts," as there is certainly no little conflict, indeed there are often fatal (immutable), irreconcilable conflicts, between the prevailing theories, as we shall see later. Only when all, or at least a large percentage, of the "counters" fall into place can the theory be transformed into a fact.

7.3 Let us imagine that our code has, say, 23 components; each component being any one of 2,000 colors. One player slots 15 of his counters into the correct place and in the right sequence. Perhaps this is a great achievement. The previous record was held by another player who achieved only 9 correct placements. Spectators are aghast. Surely this code must be the one to follow, after all it has over half of the counters in the correct place! The coding community rally round. No-one has achieved this until now; the code-breaker has to be applauded, even awarded. **The color combination he has chosen is so reminiscent of many other previous attempts**, and the owners of those attempts are only too delighted to be associated with the latest champion and to join in the applause. Perhaps some of the adoration will come his way. After all, maybe it was his choice of colors that led to the champion's eventual success!

7.4 But **the code is wrong**! A partial fit is not really a fit at all. In fact, it could be a step in the wrong direction... The rules of Mastermind state that the code-maker must award a black peg (or red peg in the above game variation) for each counter that is not only the right color but is also in the right position. He must also award a white peg for each counter which is the right color, but is in the wrong position. So, how does the code-breaker know which counters have been awarded the black pegs, which have been awarded the white pegs, and which have achieved no reward at all? He first of all recognizes a pattern, then proceeds to postulate a theory on whether this pattern will result in a 23 black peg score. On his first attempt, he is only awarded 7 black pegs and 8 white pegs. But the pattern he has in mind appears to be vindicated by the presence of the 7 black awards. He re-hashes his code, based on his theory and his thread of logic, and places his colored counters down. Now he gets 6 black pegs and 12 white ones. Is his theory correct? He has now lost a valued black award peg—one of the colors he removed on this attempt must have been correct on the previous attempt. But he has gained four somewhat valuable white pegs. Does he continue with his pattern and stubbornly insist that it will be vindicated by future attempts? Or does he admit his mistake and postulate a completely different alternative pattern? Likewise, does the scientist produce an alternative theory, or stick to his original theory come what may?

7.5 Something can be regarded as scientific evidence when a theory or hypothesis is tested objectively by someone other than the one who developed and followed through on the hypothesis, such as in an experiment or in a controlled environment. Would a carefully inscribed, etched rock encountered by travelers on a remote island be regarded as "scientific evidence" that something more than random nature caused the etching? Maybe a forensic scientist would answer "no" without being able to use his or her usual scientific instruments and methodology on the rock and perhaps to confirm that it has not been transported from another location. Let's assume that this rock projects up from deep underground and is confirmed to be very much a permanent part of this island. Now what if we were to leave out the word "scientific" and simply ask, "Does the existence of the etching in this rock give evidence that there are other forces at work here than our basic instruments can detect?" For practical purposes, our intrepid visitors would unhesitatingly answer "yes."

7.6 Similarly, let us suppose a traveler has been given a map, along with some notes indicating the path he must take on his impending journey. Some landmarks along the route are described in the order they are expected to be encountered: a large warehouse with a chimney at the back and a huge car park at the front, a man-made lake in the middle of a park, a building with a foundation stone bearing wording written at the time of the building's construction, and so on. As our traveler encounters each of these landmarks and compares them with the bequeathed instructions and the route marked out on his map, would you say that he was happy he was going in the right direction and that he had full confidence that he will indeed eventually arrive at his intended destination?

7.7 The traveler, it seems to me, is very unlike the Darwinian proponents referred to earlier who expend so much mental effort searching for pieces of the Darwin puzzle. Like the Mastermind code-breaker, there are logical indications that point in another direction; indications, evidence, that his current line of reasoning is not going to result in the complete answers he or she expects.

7.8 Einstein once said:

> "Time and again the passion for understanding has led to the illusion that man is able to comprehend the objective world rationally by pure thought without any empirical foundations—in short, by metaphysics."

7.9 To return to the Mastermind illustration, then: the attempted code of our intrepid code-breaker was wrong. Not only was the entire code "wrong," but **the pattern that he had in mind when placing his colored counters down was a long way from the pattern the code-maker had set out**. You may remember he received 7 black pegs, telling him that seven of his counters were not only of the correct color, but were also in the correct location. Of course, you see, he saw a sequence of colors, a pattern, and assumed the 7 correct counters corresponded with, and vindicated, his initial thesis. But they didn't. The black peg awards were for other counters, randomly placed within the code! ... **This is the "Mastermind Effect."**

7.10 This example aptly fits the theories associated with Darwin's survival of the fittest. If a theory is put forward that, after careful analysis and thorough testing, proves to be a fact and becomes generally accepted into the science curriculum, it usually remains unchallenged thereafter. **The Mastermind Effect, in this case, would result in a full complement of 23 black pegs**—23 perfect counters of the correct color and in the correct place. No-one could come forward with an alternative theory that proved to have a better fit!

7.11 Any inferior theory that was put alongside it now would be missing one or more reward pegs. Are there any theories connected with a Darwinian model of the universe and of the origin of life that fit this description? The illustration on the previous page shows how some textbooks, explaining Darwinian Evolution, **do not always portray an accurate picture of the relative size of skull fossils**. The set of images on the right-hand column show quite striking differences in size; but the textbook examples endeavor to indicate a similarity of form and therefore of heritage by altering the relative skull sizes.

Observe the illustration on the following page:

Is It Time to Abandon Darwin? PART I

AS SHOWN IN SOME TEXTBOOKS

REAL RELATIVE SIZE

7.12   As this work has thus far demonstrated, there are many gaps and "awkwardly fitting pieces" of the mosaic puzzle (§1.9). Fudging the pieces to make them fit does not improve the situation for Darwin's theory. Nor does it do any service for the cause of science. **By contrast, scientific findings, such as the fossil evidence from the "Cambrian Explosion," favor the Dynamic model**.

7.13   An article published in *National Geographic* ("Fossil Evidence," November 2004, p. 25) likened the fossil record to **"a film of evolution from which 999 of every 1,000 frames have been lost on the cutting-room floor."** Consider how the Mastermind Effect would apply here: Imagine that you found 100 frames of a feature film that originally had 100,000 frames. How would you determine the plot of the movie? Perhaps you could devise a preconceived idea. But what if only 5 of the 100 frames you found could be organized to support your preferred plot, while the other 95 frames tell a different story? Would you say that your preconceived idea of the movie was right based on just the five frames? Could it be that you placed the five frames in the order you did because it **seemed more in line with your theory**? Would it not be more reasonable to allow the other 95 frames to influence your opinion? In other words, would it not be better to follow *Dynamic* principles, rather than persist with the "poor fit" of the Darwinian model?

7.14   There are many **misgivings regarding the support for Darwin found in the fossil record**. Note, for example, the following list of quotations:

> "All paleontologists know that the fossil record contains precious little in the way of intermediate forms; transitions between major groups are characteristically abrupt. Gradualists usually extract themselves from this dilemma by invoking the extreme imperfection of the fossil record."

(Stephen J. Gould, *The Panda's Thumb*, 1980, p. 189.)

"What is missing are the many intermediate forms hypothesized by Darwin, and the continual divergence of major lineages into the morphospace between distinct adaptive types." (Robert L. Carroll, "Towards a new evolutionary synthesis," in *Trends in Evolution and Ecology* 15(1):27-32, 2000, p. 27.)

"Given the fact of evolution, one would expect the fossils to document a gradual steady change from ancestral forms to the descendants. But this is not what the paleontologist finds. Instead, he or she finds gaps in just about every phyletic series." (Ernst Mayr, Professor Emeritus, Museum of Comparative Zoology at Harvard University, *What Evolution Is*, 2001, p.14.)

"Given that evolution, according to Darwin, was in a continual state of motion ... it followed logically that the fossil record should be rife with examples of transitional forms leading from the less to more evolved... Instead of filling the gaps in the fossil record with so-called missing links, most paleontologists found themselves facing a situation in which there were only gaps in the fossil record, with no evidence of transformational evolutionary intermediates between documented fossil species." (Jeffrey H. Schwartz, *Sudden Origins*, 1999, p. 89.)

"He [Darwin] prophesied that future generations of paleontologists would fill in these gaps by diligent search....It has become abundantly clear that the fossil record will not confirm this part of Darwin's predictions. Nor is the problem a miserably poor record. The fossil record simply shows that this prediction was wrong." (Niles Eldridge, *The Myths of Human Evolution*, 1984, pp.45-46.)

"There is no need to apologize any longer for the poverty of the fossil record. In some ways it has become almost unmanageably rich, and discovery is out-pacing integration...The fossil record nevertheless continues to be composed mainly of gaps." (T. Neville George, "Fossils in Evolutionary Perspective," *Science Progress*, vol. 48, January 1960, pp. 1-3.)

"Despite the bright promise—that paleontology provides a means of 'seeing' evolution, it has presented some nasty difficulties for evolutionists the most notorious of which is the presence of 'gaps' in the fossil record. Evolution requires intermediate forms between species and paleontology does not provide them. The gaps must therefore be a contingent feature of the record." (David B. Kitts, *Paleontology and Evolutionary Theory*, Evolution, vol. 28, 1974, p. 467.)

"The most common explanation for the total lack of fossil evidence for fish evolution is that few transitional fossils have been preserved. This is an incorrect conclusion because every major fish kind known today has been found in the fossil record, indicating the completeness of the existing known fossil record." (Jerry Bergman, *The Search for Evidence Concerning the Origin of Fish,* CRSQ, vol. 47, 2011, p. 291.)

"Absence of the transitional fossils in the gaps between each group of fishes and its ancestor is repeated in standard treatises on vertebrate evolution.... This is one count in the creationists' charge that can only evoke in unison from the paleontologists a plea of nolo contendere [no contest]." (Arthur Strahler, *Science and Earth History*, 1987, p. 408.)

"It is interesting that all the cases of gradual evolution that we know about from the fossil record seem to involve smooth changes without the appearance of novel structures and functions." (C. Wills, *Genetic Variability*, 1989, p. 94-96.)

"We seem to have no choice but to invoke the rapid divergence of populations too small to leave legible fossil records." (S. M. Stanley, *The New Evolutionary Timetable: Fossils, Genes, and the Origin of Species*, 1981, p. 99.)

"Instead of finding the gradual unfolding of life, what geologists of Darwin's time, and geologists of the present day actually find is a highly uneven or jerky record; that is, species appear in the sequence very suddenly, show little or no change during their existence in the record, then abruptly go out of the record, and it is not always clear, in fact it's rarely clear, that the descendants were actually better adapted than their predecessors. In other words, biological improvement is hard to find." (David M. Raup, "Conflicts Between Darwin and Paleontology," *Bulletin, Field Museum of Natural History*, vol. 50, 1979, p. 23.)

"A persistent problem in evolutionary biology has been the absence of intermediate forms in the fossil record. Long term gradual transformations of single lineages are rare and generally involve simple size increase or trivial phenotypic effects. Typically, the record consists of successive ancestor-descendant lineages, morphologically invariant through time and unconnected by intermediates." (P. G. Williamson, *Palaeontological Documentation of Speciation in Cenozoic Molluscs from Turkana Basin*, 1982, p. 163.)

"What one actually found was nothing but discontinuities: All species are separated from each other by bridgeless gaps; intermediates between species are not observed . . . The problem was even more serious at the level of the higher categories." (E. Mayr, *Animal Species and Evolution*, 1982, p. 524.)

"The known fossil record is not, and never has been, in accord with gradualism. What is remarkable is that, through a variety of historical circumstances, even the history of opposition has been obscured . . . 'The majority of paleontologists felt their evidence simply contradicted Darwin's stress on minute, slow, and cumulative changes leading to species transformation.' . . . their story has been suppressed." (S. M. Stanley, *The New Evolutionary Timetable*, 1981, p. 71.)

"One must acknowledge that there are many, many gaps in the fossil record . . . There is no reason to think that all or most of these gaps will be bridged." (Ruse, *Is There a Limit to Our Knowledge of Evolution*, 1984, p.101.)

"We are faced more with a great leap of faith . . . that gradual progressive adaptive change underlies the general pattern of evolutionary change we see in the rocks . . . than any hard evidence." (N. Eldredge and I. Tattersall, *The Myths of Human Evolution*, 1982, p. 57.)

"Gaps between families and taxa of even higher rank could not be so easily explained as the mere artifacts of a poor fossil record." (Niles Eldredge, *Macro-Evolutionary Dynamics: Species, Niches, and Adaptive Peaks*, 1989, p.22.)

"To explain discontinuities, Simpson relied, in part, upon the classical argument of an imperfect fossil record, but concluded that such an outstanding regularity could not be entirely artificial." (Stephen J. Gould, *The Hardening of the Modern Synthesis*, 1983, p. 81.)

"The record jumps, and all the evidence shows that the record is real: the gaps we see reflect real events in life's history—not the artifact of a poor fossil record." (N. Eldredge and I. Tattersall, *The Myths of Human Evolution*, 1982, p. 59.)

"Gaps in the fossil record—particularly those parts of it that are most needed for interpreting the course of evolution—are not surprising." (G. L. Stebbins, *Darwin to DNA, Molecules to Humanity*, 1982, p. 107.)

"The fossil record itself provided no documentation of continuity—of gradual transition from one animal or plant to another of quite different form." (S. M. Stanley, *The New Evolutionary Timetable: Fossils, Genes and the Origin of Species*, 1981, p. 40.)

"The absence of fossil evidence for intermediary stages between major transitions in organic design, indeed our inability, even in our imagination, to construct functional intermediates in many cases, has been a persistent and nagging problem for gradualistic accounts of evolution." (Stephen J. Gould, *Is a New and General Theory of Evolution Emerging*?, 1982, p. 140.)

"The lack of ancestral or intermediate forms between fossil species is not a bizarre peculiarity of early metazoan history. Gaps are general and prevalent throughout the fossil record.... Gaps between higher taxonomic levels are general and large." (R. A. Raff and T. C. Kaufman, *Embryos, Genes, and Evolution: The Developmental-Genetic Basis of Evolutionary Change*, 1991, pp. 34, 35.)

"We have so many gaps in the evolutionary history of life, gaps in such key areas as the origin of the multicellular organisms, the origin of the vertebrates, not to mention the origins of most invertebrate groups." (C. McGowan, *In the Beginning . . . A Scientist Shows Why Creationists are Wrong*, 1984, p. 95.)

"If life had evolved into its wondrous profusion of creatures little by little, Dr. Eldredge argues, then one would expect to find fossils of transitional creatures which were a bit like what went before them and a bit like what came after. But no one has yet found any evidence of such transitional creatures. This oddity has been attributed to gaps in the fossil record which gradualists expected to fill when rock strata of the proper age had been found. In the last decade, however, geologists have found rock layers of all divisions of the last 500 million years and no transitional forms were contained in them. If it is not the fossil record which is incomplete then it must be the theory." (*The Guardian Weekly*, 26 Nov 1978, vol. 119, no 22, p. 1.)

> "People and advertising copywriters tend to see human evolution as a line stretching from apes to man, into which one can fit new-found fossils as easily as links in a chain. Even modern anthropologists fall into this trap . . .[W]e tend to look at those few tips of the bush we know about, connect them with lines, and make them into a linear sequence of ancestors and descendants that never was. But it should now be quite plain that the very idea of the missing link, always shaky, is now completely untenable." (Henry Gee, "Face of Yesterday," *The Guardian*, Thursday July 11, 2002.)

> "To take a line of fossils and claim that they represent a lineage is not a scientific hypothesis that can be tested, but an assertion that carries the same validity as a bedtime story—amusing, perhaps even instructive, but not scientific.... The intervals of time that separate the fossils are so huge that we cannot say anything definite about their possible connection through ancestry and descent." (Henry Gee, *In Search of Deep Time— Beyond the Fossil Record to a New History of Life*, 1999, pp. 116-117 & p. 23.)

7.15   The list of quotations from scientists on the topic of the "jerkiness" of the fossil record seems endless:

> "Instead of finding the gradual unfolding of life," said evolutionary paleontologist David M. Raup, "what geologists of Darwin's time, and geologists of the present day actually find is a **highly uneven or jerky record**; that is, species appear in the sequence very suddenly, show little or no change during their existence in the record, then abruptly go out of the record."—*Field Museum of Natural*

> *History Bulletin,* "Conflicts Between Darwin and Paleontology," by David M. Raup, January 1979, p. 23.

7.16 It is the bold prediction of this work that Dynamic Evolution will become the **replacement for many if not all of the assertions of the Darwinian Model**. Perhaps this is not as bold as you might think. For example, in an interview in 2008, evolutionary biologist Stuart Newman discussed the need for a new theory of evolution that could explain the sudden appearance of novel forms of life. He said:

> "The Darwinian mechanism that's used to explain all evolutionary change will be **relegated, I believe, to being just one of several mechanisms**—maybe not even the most important when it comes to understanding macroevolution, the evolution of major transitions in body type."— *Archaeology, The Origin of Form Was Abrupt Not Gradual,* by Suzan Mazur, October 11, 2008.

7.17 At a meeting held in Chicago in the 1980's, some 150 specialists in evolution held a four-day conference on the subject "Macroevolution." *Science,* the official journal of the **American Association for the Advancement of Science**, reported the mood: "Clashes of personality and academic sniping created palpable tension … the proceedings were at times unruly and even acrimonious." Many scientists complained that "a large proportion of the contributions were characterized more by description and assertion than by the presentation of data."

7.18 "The pattern that we were told to find for the last 120 years does not exist," declared Niles Eldridge, paleontologist from the American Museum of Natural History in New York. He believes new species arise, not from gradual changes, but in **sudden bursts of evolution, in harmony with the principles of Dynamic Evolution**.

7.19 At the same conference, Stephen Jay Gould of Harvard said: "Certainly the record is poor, but the jerkiness you see is not the result of gaps, it is the consequence of the jerky mode of evolutionary change." Everett Olson, a paleontologist, remarked: **"I take a dim view of the fossil record as a source of data**." Francisco Ayala, formerly an advocate of Darwinian Evolution, added: "I am now convinced from what the paleontologists say that **small changes do not accumulate**."

7.20 *Science* magazine summed up the controversy:

> "The central question of the Chicago conference was whether the mechanisms underlying microevolution [small changes] can be extrapolated to explain the phenomena of macroevolution [big jumps across species boundaries]. . . . the answer can be given as a clear, No."

7.21 One example found in nature that provides support for Dynamic Evolution, whilst at the same time presenting a thorny problem for Darwinian advocates, is the octopus. **The octopus is no relative of Homo sapiens, but his eye is amazingly "human."** Unrelated fish and eels have electrical shocking equipment. Unrelated insects, worms, bacteria, and fish have luminous organs giving off cold light. Unrelated lampreys, mosquitoes, and leeches have anticoagulants to keep their victims' blood from clotting. Unrelated porcupines, echidnas, and hedgehogs are said to have independently evolved quills. Unrelated dolphins and bats have sonar systems. Unrelated fish and insects have bifocal eyes for vision in air and under water. In many unrelated animals—crustaceans, fish, eels, insects, birds, mammals—there are "dynamic" abilities for migration.

7.22 Darwinian evolutionists support the idea that, along **three different (independent) lines, warm-blooded animals developed from cold-blooded reptiles; three times color vision developed independently; five times wings and flight developed by Darwinian principles in unrelated fish, insects, pterodactyls, birds, and mammals**. However, it is clear from the argumentation presented in this volume that Dynamic Evolution provides the mechanism for these devices and that the basis for the principles is superior to the Darwinian support.

7.23 In later studies, the "missing pieces" of Darwinian Evolution are compared alongside the comprehensive explanations of the Dynamic model.

Part II provides a comprehensive, systematic study of the principles outlined in Part I. In a series of diagrams, and cross-referenced, "compartmentalized" analyses, the seamless connections between Entities is demonstrated. Principles that reach many levels of the Plexus are given particular attention, as are the relationships that form a repeating pattern throughout.

Part III finalizes the argumentation of the first two volumes, enabling a complete understanding of all the principles that make Dynamic Evolution the "most convincing and attractive alternative to the theory of Darwinian Evolution to date." (§1.19)

Appendices

8 Glossary

alpha particle (α)	A helium nucleus (two protons and two neutrons) ejected from an unstable nucleus. Alpha particles are stopped by a few centimeters of air, or thin paper. They exhibit strong ionizing radiation, due partly to relatively large mass and double positive charge—about 8,000 times greater than beta particles.
atom	The smallest unit of matter that can exist independently. Atoms are made up of a nucleus consisting of (positively charged) protons and (uncharged) neutrons, and a number of (negatively charged) electrons that orbit the nucleus. An electrically neutral atom has as many protons as it has electrons.
beta particle	A high speed electron emitted in a form of radioactive disintegration from an atomic nucleus which has an excess of electric charge and energy. They have less mass and are less charged than alpha particles and interact less intensely with atoms in the materials that they pass through, which gives them a longer range than alpha particles.
boson	A particle with a symmetric wave function. Two bosons can occupy the same quantum state.

consequential entities	Everything that occurred and that came into existence after the Big Bang; the facets, or properties, of forces, conditions, and objects in the universe that are subject to certain consistent patterns of nature, performance, or form, that vary little or not at all from the laws that govern their structure and behavior.
conservation of energy	The law that the sum of all the energies of the universe, actual and potential, is unchangeable.
conservation of momentum	The principle stating that for any isolated system, linear momentum is constant with time.
constructive interference	The amplification of one wave by another, identical wave, of the same sign. Two constructively interfering waves are said to be "in phase."
dependencies	Entities within the Plexus that require the existence, continuity, and functionality of other Entities.
destructive interference	The cancellation of one wave by another wave that is exactly out of phase with the first. Despite the dramatic name of this phenomenon, nothing is "destroyed" by this interference—the two waves emerge intact once they have passed each other.

diffraction	The deviation of light from rectilinear propagation, is a characteristic of wave phenomena which occurs when a portion of a wave front is obstructed in some way. When various portions of a wave front propagate past some obstacle, and interfere at a later point past the obstacle, the pattern formed is called a diffraction pattern.
dispersion	The dependence of the velocity of a wave on the frequency of the wave is known as dispersion. A media in which waves of different frequencies propagate at different speeds is said to be dispersive.
dynamics	The application of kinematics to understand why objects move the way they do. More precisely, dynamics is the study of how forces cause motion.
electricity	The flow of electric charge. It has a voltage (energy), may cause a current (flow) and can be slowed down or cancelled by resistance.
electromagnetic spectrum	The spectrum containing all the different kinds of electromagnetic waves, ranging in wavelength and frequency.
electromagnetic wave	Transverse traveling wave created by the oscillations of an electric field and a magnetic field. Electromagnetic waves travel at the

	speed of light, m/s. Examples include microwaves, X rays, and visible light.
electron	Fundamental particle possessing both charge and spin. They are approximately 1/2000 of a proton or neutron. Electrons are responsible for the bonding that forms solids from free atoms, delocalized electrons that give rise to the conducting properties of metals. The fact that the electron possesses spin produces the magnetic properties of materials, along with the orbital motion of the electrons around and between atoms.
energy	A property of something that dictates its potential for change; a conserved scalar quantity associated with the state or condition of an object or system of objects. Types of energy include: kinetic, potential, thermal, chemical, mechanical, and electrical.
exchange	The interaction that gives rise to magnetic ordering; comes from the rules determining how indistinguishable particles, such as electrons, can be exchanged between systems in quantum mechanics.
fermat's principle	The principle of least time states that in traveling from point A to point B light follows the path which takes the least time.

fermion	A particle that follows Pauli's principle, where no two fermions may have the same quantum state.
fine-tuning	Consequential Entities that have specific parameters (sizes, speeds, strengths, distances, etc.) that only work effectively (or at all) if those parameters are within a particular value threshold; for example, plus or minus 1% (+-1%).
force	A push or a pull that causes an object to accelerate. Newton's 2nd law defines a force as being proportional to the acceleration it produces.
frequency	The number of oscillations or vibrations per unit of time. SI units Hz (Hertz). Frequency is the HE inverse of the period T.
galaxy	A combined group of millions of stars held together by gravity.
group velocity	The speed of the modulated signal. $v_g = dw/dk$
Heisenberg principle	Also called the 'uncertainty principle,' states that the position and velocity of an object cannot both be measured exactly, at the same time, even theoretically. The concepts of exact *position* and exact *velocity* together have no meaning in physics.

huygens' principle	Each point on a propagating wave is considered as the source of a small wave, or wavelet which propagates at the same speed as the original wave. The new wave front for the propagating wave is constructed by drawing a line tangent to the edges of these wavelets.
index of refraction	The ratio of the speed of an electromagnetic wave in vacuum to the speed of the electromagnetic wave in a particular medium, typically denoted by the lowercase n. $n = c/v$. The index of refraction is a property of the medium in question, and the incident electromagnetic wave. In general, the index of refraction may depend on the frequency of the incident wave, and on its polarization. A media in which the index of refraction depends on the frequency of the electromagnetic wave is said to be dispersive. A media in which the index of refraction depends on the polarization of the incident wave is said to be birefringent.
inertia	The tendency of an object to remain at a constant velocity, or its resistance to being accelerated (=Newton's First Law).
inertial reference frame	A reference frame in which Newton's First Law is true. Two inertial reference frames move at a constant

	velocity relative to one another. According to Einstein's theory of special relativity, the laws of physics are the same in all inertial reference frames.
irradiance	The intensity of a light wave; the power per unit area carried by an electromagnetic wave. The time averaged value of the magnitude of the Poynting vector.
Kepler's first law	The path of each planet around the sun is an ellipse with the sun at one focus.
Kepler's second law	If a line is drawn from the sun to the planet, then the area swept out by this line in any given time interval is constant.
Kepler's third law	Given the period, T, and semi-major axis, a, of a planet's orbit, the ratio is the same for every planet.
law of conservation of energy	Energy cannot be made or destroyed; energy can only be changed from one place to another or from one form to another
mass	A property that is equivalent to the number of atoms or the amount of energy that something contains.
meson	A class of elementary particle whose mass is between that of a proton and that of an electron.

moment	A magnetic dipole. Moments will align with a uniform field due to the torque exerted but will experience a force in a field gradient.
monochromatic	Light of a single wavelength.
neutron	A neutrally charged particle that, along with protons, constitutes the nucleus of an atom.
neutron number	The number, N, of neutrons in an atomic nucleus.
nucleus	The "lumpy" center of the atom, consisting of protons and neutrons bound together by the strong nuclear force.
phase velocity	The velocity of a carrier wave in a modulated signal.
photoelectric effect	When electromagnetic radiation shines on metal, the surface of the metal releases energized electrons. The way in which these electrons are released supports the quantum view according to which electromagnetic waves are treated as particles.
photon	Light manifested as a particle.
plexus	An illustrative object (model) that demonstrates the nature, qualities, inter-connectivity, and form of the universal Consequential Entities and contains associated interactive principles that serve to explain the

	development and sustainability of those Entities.
proton	A positively charged particle that, along with the neutron, occupies the nucleus of the atom.
quark	The building blocks of all matter, and the constituent parts of protons, neutrons, and mesons. Three quarks combine to make both protons and neutrons. Items that are made of quarks are called hadrons.
scalar	A quantity that possesses a magnitude but not a direction. Mass and length are common examples.
specific heat	The amount of heat of a material required to raise the temperature of either one kilogram or one gram of that material by one degree Celsius.
spin	The intrinsic angular momentum possessed by fundamental particles—giving the appearance of "spinning." This angular momentum leads to a magnetic moment of fixed size for each particle.
strong nuclear force	The force that binds protons and neutrons together (by interaction with the constituent quarks) in the atomic nucleus.
temperature	A measure of the average kinetic energy of the molecules in a system.

	Temperature is related to heat by the specific heat of a given substance.
uncertainty principle	A principle derived by Werner Heisenberg in 1927 that states that we can never determine both the position and the momentum of a particle as a single experiment.
wave	A system with many parts in periodic, or repetitive, motion. The oscillations in one part cause vibrations in nearby parts.
wave speed	The speed at which a wave crest or trough propagates. Note that this is not the speed at which the actual medium moves.
wavelength	The distance between successive wave crests, or troughs, measured in meters, and is related to frequency and wave speed by $\lambda = v/f$ (Wavelength = Speed / Frequency).
weak nuclear force	The force that produces beta decay and changes a proton to a neutron and releases an electron and a neutrino. Change from proton to neutron (and vice versa) occurs with the switch of up quarks to down quarks and down quarks to up quarks.

9 BIBLIOGRAPHY

Agre, P., M. Bonhivers, and M. J. Borgnia. (1998).The aquaporins, blueprints for cellular plumbing systems. Journal of Biolgical Chemistry, 273, 14659–14662

Alioto, Anthony M., A History of Western Science, second edition, Englewood Cliffs, NJ: Prentice-Hall, 1993.

American Public Health Association. Standard methods for the examination of water and wastewater, 17th ed. Washington, DC, 1989.

Aristotle, On the Heavens [De Caelo], translated by W. K. C. Guthrie, Cambridge, MA: Harvard University Press, 1960.

Bederson, Benjamin (editor), More Things in Heaven and Earth: A Celebration of Physics at the Millennium, New York: Springer Verlag, 1999. A collection of review articles, some historical.

Ben-David, Joseph, The Scientist's Role in Society: A Comparative Study, Chicago: University of Chicago Press, 1984.

Bensaude-Vincent, Bernadette and Stengers, Isabelle, A History of Chemistry, Cambridge, MA: Harvard University Press, 1996.

Blay, Michel, Reasoning with the Infinite: From the Closed World to the Mathematical Universe. Chicago: University of Chicago Press, 1998.

Boas, Marie, see Hall, Marie Boas

Boorse, H. A. and Motz, L. (editors), The World of the Atom. New York: Basic Books, 1966.

Borgnia, M., S. Nielsen, A. Engel, and P. Agre. (1999). Cellular and molecular biology of the aquaporin water channels. Annual Review of Biochemistry, 68, 425–458.

Bridgman, P. W., The Logic of Modern Physics, New York: Macmillan, 1960.

Brock, William H., The Norton History of Chemistry, New York: Norton, 1993

Brown, Laurie M., Pais, Abraham, and Pippard, Brian (editors), Twentieth Century Physics, Philadelphia: Institute of Physics Publishing, 1995.

Brush, Stephen G. (editor), History of Physics: Selected Reprints, College Park, MD: American Association of Physics Teachers, 1988.

Brush, Stephen G. (editor), Kinetic Theory, Vol. 1, The Nature of Gases and of Heat; Vol. 2, Irreversible Processes; Vol. 3, The Chapman-Enskog Solution of the Transport Equation for Moderately Dense Gases, New York: Pergamon Press, 1965-1972

Brush, Stephen G. (editor), Resources for the History of Physics, Hanover, N.H.: University Press of New England, 1972.

Brush, Stephen G., and Belloni, L., The History of Modern Physics: An International Bibliography, New York: Garland, 1983.

Brush, Stephen G., Statistical Physics and the Atomic Theory of Matter, from Boyle and Newton to Landau and Onsager, Princeton, NJ: Princeton University Press, 1983

Brush, Stephen G., The Kind of Motion we call Heat: A History of the Kinetic Theory of Gases in the 19th Century, Amsterdam: North-Holland, 1986

Buchwald, Jed Z. (editor), Scientific Practice: Theories and Stories of Doing Physics, Chicago: University of Chicago Press, 1995

Burykin and A. Warshel (2003). What really prevents proton transport through aquaporin? Charge self-energy vs. proton wire proposals, Biophysical Journal 85, 3696-3706

Butterfield, Herbert, Origins of Modern Science, 1300-1800. Revised edition. New York: Free Press, 1997.

Cardwell, D. S. L., From Watt to Clausius: The Rise of Thermodynamics in the Early Industrial Age, Ames, IA: Iowa State University Press, 1989

Cellular Microbiology. 2012

Chakrabarti, N., Tajkhorshid, E., Roux, B. and Pommes, R. (2004). Molecular basis of proton blockage in aquaporins, Structure 12, 65-74

Cohen, I. Bernard and Westfall, Richard S. (editors), Newton: Texts, Backgrounds, Commentaries, New York: Norton, 1995.

Cohen, I. Bernard, An Introduction to Newton's Principia, Cambridge, MA: Harvard University Press, 1971

Cohen, I. Bernard, The Birth of a New Physics, revised & updated edition, New York: Norton, new edition, 1985.

Cohen, I. Bernard, The Newtonian Revolution, With Illustrations of the Transformation of Scientific Ideas, New York: Cambridge University Press, 1980

Collingwood, R. G., The Idea of Nature, Oxford: Clarendon Press, 1964.

Conant, J. B. (editor), Harvard Case Histories in Experimental Science, Cambridge, MA: Harvard University Press, 1966

Conant, J. B., On Understanding Science, New York: New American Library, 1951.

Conant, J. B., Science and Common Sense, New Haven, CT: Yale University Press, 1961

Crease, Robert P., and C. Mann, Charles C., The Second Creation: Makers of the Revolution in 20th-century Physics, revised edition, New Brunswick, NJ: Rutgers University Press, 1996.

Crombie, A. C., Medieval and Early Modern Science, Cambridge, MA: Harvard University Press, 1967.

Crombie, A. C., Science, Art and Nature in Medieval and Modern Thought, London: Hambledon Press, 1996.

Crowe, Michael J., Modern Theories of the Universe from Herschel to Hubble, New York: Dover, 1994.

Crowe, Michael J., Theories of the World from Antiquity to the Copernican Revolution, New York: Dover, 1990.

Cushing, J. T., Philosophical Concepts in Physics: The Historical Relation between Philosophy and Scientific Theories, New York: Cambridge University Press, 1998

Davis, E. A. (editor), Science in the Making: Scientific Development as Chronicled by Historic Papers in the Philosophical Magazine, Levittown, PA: Taylor & Francis, 1995-1999. Four volumes covering 1798-1998.

de Groot, B. L., and Grubmüller, H. (2001). Water permeation across biological membranes: mechanism and dynamics of aquaporin-1 and GlpF, Science 294, 2353-2357

de Groot, B. L., Frigato, T., Helms, V. and Grubmüller, H. (2003). The mechanism of proton exclusion in the aquaporin-1 channel, Journal of Molecular Biology 333, 279-293

Dear, Peter (editor), The Scientific Enterprise in Early Modern Europe: Readings from Isis, Chicago: University of Chicago Press, 1997.

Densmore, Dana, Newton's Principia: The Central Argument, translation, notes & expanded proofs by Dana Densmore; translation and illustrations by W. H. Donahue, Santa Fe, NM: Green Lion Press, 1995.

DeVorkin, David, The History of Modern Astronomy and Astrophysics: A Selected, annotated Bibliography, New York: Garland, 1982.

Dijksterhuis, E. J., The Mechanization of the World Picture, translated by C. Dikshoorn, New York: Oxford University Press, 1969

Drake, Stillman (editor and translator), Discoveries and Opinions of Galileo, Garden City, NY: Doubleday Anchor Books, 1957

Drake, Stillman, Galileo, New York: Hill and Wang, 1980

Dugas, René, History of Mechanics, New York: Dover, 1988.

Dugas, René, Mechanics in the 17th Century, New York: Central Book Co., 1958.

Duhem, Pierre, Aim and Structure of Physical Theory, translated by P. P. Wiener, New York: Atheneum, 1962.

Einstein, Albert, "Autobiographical Notes," in Albert Einstein Philosopher-Scientist (edited by P. A. Schilpp), pages 1-95, La Salle, IL: Open Court, 1970; also reprinted as a separate book, Autobiographical Notes: A Centennial Edition, Chicago: Open Court, 1991

Einstein, Albert, and Infeld, Leopold, The Evolution of Physics, from early concepts to Relativity and Quanta, New York: Simon and Schuster, 1966

Einstein, Albert, Collected Papers, edited by John Stachel et al., Princeton, NJ: Princeton University Press, 1987-

Einstein, Albert, Ideas and Opinions, translated by Sonja Bargmann, New York: Modern Library, 1994.

Einstein, Albert, Relativity: The Special and the General Theory, translated by R. W. Lawson, second edition, New York: Crown, 1995.

Everdell, William R., The First Moderns: Profiles in the Origins of 20th-Century Thought, Chicago: University of Chicago Press, 1997

Everitt, C. W. F., James Clerk Maxwell, Physicist and Natural Philosopher, New York: Scribner, 1975

Feynman, Richard P., The Character of Physical Law, Cambridge, MA: MIT Press, 1973.

Frank, Phillipp, Einstein, His Life and Times, New York: Da Capo Press, 1979

French, A. P. and Greenslade, T. B., Jr. (editors), Physics History from AAPT Journals, II, College Park, MD: American Association of Physics Teachers, 1995.

Freund, Ida, The Study of Chemical Composition, An Account of its Method and Historical Development, New York: Dover, 1968

Fu, D., Libson, A., Miercke, L. J., Weitzman, C., Nollert, P., Krucinski, J., and Stroud, R. M. (2000). Structure of a glycerol-conducting channel and the basis for its selectivity, Science 290, 481-6

Galileo Galilei, Dialogue Concerning the Two Chief World Systems, translated with revised notes by S. Drake, Berkeley, CA: University of California Press, 1967

Galileo Galilei, Discoveries and Opinions of Galileo, translated with notes by S. Drake, Garden City, NY: Doubleday Anchor Books, 1957

Galileo Galilei, Two New Sciences, Including Centers of Gravity & Force of Percussion, translated by S. Drake, second edition, Toronto: Wall & Thompson, 1989

Galison, Peter, How Experiments End, Chicago: University of Chicago Press, 1987

Galison, Peter, Image and Logic: A Historical Culture of Microphysics, Chicago: University of Chicago Press, 1997.

Garber, Elizabeth, Brush, Stephen G., and Everitt, C. W. F. (editors), Maxwell on Molecules and Gases, Cambridge, MA: MIT Press, 1986

Garber, Elizabeth, The Language of Physics: The Calculus and the Development of Theoretical Physics in Europe, 1750-1914, Boston: Birkhäuser, 1999

Geymonat, Ludovico, Galileo Galilei, A Biography and Inquiry into his Philosophy of Science, translated by S. Drake, New York: McGraw-Hill, 1965

Gillispie, C. C. (editor), Dictionary of Scientific Biography, Vols. 1-16, New York: Scribner, 1970-80; supplements, Vols. 17-18 (1990) edited by F. L. Holmes

Gjertsen, D., The Newton Handbook, London: Routledge and Kegan Paul, 1986

Grant, Edward, The Foundations of Modern Science in the Middle Ages: Their Religious, Institutional, and Intellectual Contexts, New York: Cambridge University Press, 1996

Hall, A. Rupert, and Hall, Marie Boas, A Brief History of Science, Ames: Iowa State University Press, 1988

Hall, A. Rupert, From Galileo to Newton, 1630-1720, New York: Dover, 1981

Hall, A. Rupert, The Revolution in Science 1500-1750, New York: Longman, 1983

Hall, Marie Boas (editor), Nature and Nature's Laws: Documents of the Scientific Revolution, New York: Harper & Row, 1970

Hall, Marie Boas, The Scientific Renaissance 1450-1630, New York: Harper, 1966

Hanson, Norwood Russell, Constellations and Conjectures, Boston: Reidel, 1973

Harman, P. M., Energy, Force, and Matter: The Conceptual Development of Nineteenth-Century Physics, New York: Cambridge University Press, 1982

Heilbron, J. L., and B. R. Wheaton, B. R., Literature on the History of Physics in the 20th Century, Berkeley, CA: University of California, Office for the History of Science and Technology, 1981

Heilbron, J. L., Elements of Early Modern Physics, Berkeley: University of California Press, 1982

Henry, John, The Scientific Revolution and the Origins of Modern Science, New York: St. Martin's Press, 1997

Hetherington, Norriss S. (editor), Cosmology: Historical, Literary, Philosophical, Religious, and Scientific Perspectives, New York: Garland, 1993

Hetherington, Norriss S. (editor), Encyclopedia of Cosmology: Historical, Philosophical, and Scientific Foundations of Modern Cosmology, New York: Garland, 1993.

Hoffmann, Dieter, Bevilacqua, Fabio, and Stuewer, Roger H. (editors), The Emergence of Modern Physics, Pavia, Italy: Universita degli Studi di Pavia, 1996.

Holmes, F. L. (editor), Dictionary of Scientific Biography, Supplement II (Vols. 17 & 18), New York: Scribner, 1990, includes biographies of scientists who died since 1970.

Holton, Gerald, Einstein, History, and other Passions: The Rebellion against Science at the End of the Twentieth Century, Reading, MA: Addison-Wesley, 1996.

Holton, Gerald, The Advancement of Science and its Burdens, Cambridge, MA: Harvard University Press, 1998.

Holton, Gerald, The Scientific Imagination, Cambridge, MA: Harvard University Press, 1998.

Holton, Gerald, Thematic Origins of Scientific Thought, Kepler to Einstein, revised edition, Cambridge, MA: Harvard University Press, 1988.

Hoskin, Michael (editor), The Cambridge Illustrated History of Astronomy, New York: Cambridge University Press, 1997.

Howson, Colin (editor), Method and Appraisal in the Physical Sciences, New York: Cambridge University Press, 1976.

Ihde, Aaron J., The Development of Modern Chemistry, New York: Harper & Row, 1964.

Ilan, B., Tajkhorshid, E., Schulten, K. and Voth, G. (2004). The mechanism of proton exclusion in aquaporin water channels. PROTEINS: Structure, Function, and Bioinformatics, 55, 223-228

International Organization for Standardization. Water quality series. Geneva. Rodier J. L'analyse de l'eau. Eaux naturelles, eaux résiduaires, eau de mer. 7th ed. Paris, Dunod, 1984.

Jacob, James R., The Scientific Revolution: Aspirations and Achievements, 1500-1700, Amherst, NY: Humanity Books, 1999.

Jammer, Max, Conceptual Development of Quantum Mechanics, second edition, New York: Tomash/American Institute of Physics, 1989.

Jensen, M. O., Tajkhorshid, E., and Schulten, K. (2003). Electrostatic tuning of permeation and selectivity in aquaporin water channels, Biophysical Journal 85, 2884-2899

Jungnickel, Christa, and McCormmach, Russell Intellectual Mastery of Nature, Vol. I, The Torch of Mathematics 1800-1870; Vol. 2, The Now Mighty Theoretical Physics 1870-1925, Chicago: University of Chicago Press, 1986.

Kepler, Johannes, The Harmony of the World, translated with introduction and notes by E. J. Aiton, A. M. Duncan and J. V. Field, Philadelphia: American Philosophical Society, 1997 (Memoirs, Vol. 209)

Kevles, Daniel J., The Physicists: The History of a Scientific Community in Modern America, Cambridge, MA: Harvard University Press, 1995.

Kitcher, Philip, The Advancement of Science: Science without Legend, Objectivity without Illusions, New York: Oxford University Press, 1993.

Knight, David, Ideas in Chemistry, New Brunswick, NJ: Rutgers University Press, 1992.

Koyré, Alexandre, From the Closed World to the Infinite Universe, Baltimore, MD: Johns Hopkins University Press, 1968.

Koyré, Alexandre, Metaphysics and Measurement, New York: Gordon & Breach, 1992.

Koyré, Alexandre, Newtonian Studies, Cambridge, MA: Harvard University Press, 1965.

Krige, John, and Pestre, Dominique (editors), Science in the 20th Century, Amsterdam: Harwood Academic, 1997.

Kuhn, Thomas S., The Copernican Revolution, New York: Fine Communications, 1997.

Kuhn, Thomas S., The Essential Tension: Selected Studies in Scientific Tradition and Change, Chicago: University of Chicago Press, 1977.

Kuhn, Thomas S., The Structure of Scientific Revolutions, third edition, Chicago: University of Chicago Press, 1996.

Lang, Kenneth R., and Gingerich, Owen (editors), Source Book in Astronomy and Astrophysics, 1900-1975, Cambridge, MA: Harvard University Press, 1979.

Lankford, John (editor), History of Astronomy: An Encyclopedia, New York: Garland, 1997.

Leverington, David, A History of Astronomy, from 1890 to the present, New York: Springer-Verlag, 1995.

Liese, Jan; Rooijakkers, Suzan H M; van Strijp, Jos A G; Novick, Richard P; Dustin, Michael L

Lightman, Alan, Great Ideas in Physics, second edition, New York: McGraw-Hill, 1997.

Lindberg, David, The Beginnings of Western Science, Chicago: University of Chicago Press, 1992.

Lindberg, David, Theories of Vision from al-Kindi to Kepler, Chicago: University of Chicago Press, 1976.

Lodge, Oliver, Pioneers of Science, and the Development of their Scientific Theories, New York: Dover, 1960.

Lucretius, On the Nature of the Universe, translated by Ronald Melville from De Rerum Natura, with introduction and notes by Don and Peta Fowler, Oxford: Clarendon Press, 1997.

Mach, Ernst, Popular Scientific Lectures, translated by T. J. McCormack, Chicago: Open Court, 1986.

Mach, Ernst, The Science of Mechanics, translated by T. J. McCormack, sixth edition, LaSalle, IL: Open Court, 1960.

Machamer, Peter (editor), The Cambridge Companion to Galileo, New York: Cambridge University Press, 1998.

Mason, Stephen F., A History of the Sciences, new revised edition, New York: Collier Books, 1962.

Mauskopf, Seymour H. (editor), Chemical Sciences in the Modern World, Philadelphia: University of Pennsylvania Press, 1993.

Maxwell, James Clerk, Maxwell on Molecules and Gases (edited by E. Garber et al.), Cambridge, MA: MIT Press, 1986.

Maxwell, James Clerk, The Scientific Letters and Papers (edited by P. M. Harman), New York: Cambridge University Press, 1990-

McClellan, James E., III, and Dorn, Harold, Science and Technology in World History: An Introduction, Baltimore: Johns Hopkins University Press, 1999.

McKenzie, A. E. E., The Major Achievements of Science, Ames: Iowa State University Press, 1988.

Merton, Robert K., "Singletons and Multiples in Scientific Discovery," Proceedings of the American Philosophical Society, Vol. 57, pages 1-23 (1969)

Merton, Robert K., The Sociology of Science, Chicago: University of Chicago Press, 1973.

Moyer, Albert E., "History of Physics" [in 20th century America], Osiris, new series, Vol. 2, pages 163-182 (1985)

Moyer, Albert E., American Physics in Transition, Los Angeles: Tomash, 1983.

Murata, K., Mitsuoka, K., Hirai, T., Walz, T., Agre, P., Heymann, J. B., Engel, A., and Fujiyoshi, Y. (2000). Structural determinants of water permeation through aquaporin-1, Nature 407, 599-605

Newton, Isaac, Newton's Philosophy of Nature, selections from his writings, edited and arranged with notes by H. S. Thayer, New York: Hafner, 1953.

Newton, Isaac, Opticks: Or, A Treatise of the Reflections, Refractions, Inflections & Colours of Light, based on the fourth edition (1730), New York: Dover, 1979.

Newton, Isaac, The Principia: Mathematical Principles of Natural Philosophy, translated by I. B. Cohen and Anne Whitman, with the assistance of Julia Budenz, preceded by a Guide to Newton's Principia by I. B. Cohen, Berkeley: University of California Press, 1999.

North, John, Norton History of Astronomy and Cosmology, New York: Norton, 1995.

Nye, Mary Jo, Before Big Science: The Pursuit of Modern Chemistry and Physics 1800-1940, Cambridge, MA: Harvard University Press, 1999.

Nye, Mary Jo, From Chemical Philosophy to Theoretical Chemistry: Dynamics of Matter and Dynamics of Disciplines, 1800-1950, Berkeley, CA: University of California Press, 1993.

Olby, R. C., Cantor, G. N., Christie, J. R. R., and Hodge, M. J. S. (editors), Companion to the History of Modern Science, New York: Routledge, 1996.

Pauli, Wolfgang, et al. (editors), Niels Bohr and the Development of Physics, New York: McGraw-Hill, 1965.

Paustian T, Roberts G (2009). Through the Microscope: A Look at All Things Small (3rd ed.). Textbook Consortia.

Planck, Max, Scientific Autobiography and other Papers, translated by F. Gaynor, New York: Greenwood Press, 1968.

Plato, Phaedon, translated by D. Gallop, New York: Oxford University Press, 1993.

Poincaré, Henri, Science and Hypothesis, New York: Dover, 1952.

Poincaré, Henri, Science and Method, translated by F. Maitland, New York: Dover, 1952.

Poincaré, Henry, The Value of Science, translated by G. B. Halsted, New York: Dover, 1958.

Popper, Karl, Conjectures and Refutations: The Growth of Scientific Knowledge, second edition, New York: Basic Books, 1965.

Popper, Karl, Logic of Scientific Discovery, New York: Basic Books, 1959.

Preston, G. M., P. Piazza-Carroll, W. B. Guggino, and P. Agre. (1992). Appearance of water channels in Xenopus oocytes expressing red cell CHIP28 water channel. Science, 256, 385–387

Purrington, Robert D., Physics in the Nineteenth Century, New Brunswick, NJ: Rutgers University Press, 1997.

Pyenson, Lewis, and Sheets-Pyenson, Susan, Servants of Nature: A History of Scientific Institutions, Enterprises and Sensibilities. New York: Norton, 1999.

Rayner-Canham, Marelene, and Rayner-Canham, Geoffrey W. (editors), A Devotion to their Science: Pioneer Women of Radioactivity, Philadelphia: Chemical Heritage Foundation, 1997.

Ren, G., Reddy, V. S., Cheng, A., Melnyk, P., and Mitra, A. K. (2001).Visualization of a water-selective pore by electron crystallography in vitreous ice, Proc Natl Acad Sci U S A 98, 1398-1403

Rigden, John S. (editor), Macmillan Encyclopedia of Physics, four volumes, New York: Macmillan Reference USA/Simon & Schuster Macmillan, 1996.

Ronan, Colin A., Science: Its History and Development among the World's Cultures, New York: Facts on File, 1982.

Sambursky, S., Physical Thought from the Presocratics to the Quantum Physicists: An Anthology, New York: Pica, 1975.

Sarton, George, The Study of the History of Science, New York: Dover, 1957.

Schilpp, P. A. (editor), Albert Einstein Philosopher-Scientist, La Salle, IL: Open Court, 1970.

Schirmacher, Wolfgang (editor), German Essays on Science in the 20th Century, New York: Continuum, 1996.

Schlagel, Richard H., From Myth to Modern Mind: A Study of the Origins and Growth of Scientific Thought, Vol. 1, Theogony through Ptolemy, Vol. 2, Copernicus through Quantum Mechanics, New York: Peter Lang, 1996.

Schneer, Cecil J., Mind and Matter: Man's Changing Concepts of the Material World, Ames, IA: Iowa State University Press, 1988.

Schneer, Cecil J., The Evolution of Physical Science, Lanham, MD: University Press of America, 1984.

Schorn, Ronald A., Planetary Astronomy: From Ancient Times to the Third Millenium, College Station, TX: Texas A & M University Press, 1998.

Segrè, Emilio, From X-Rays to Quarks: Modern Physicists and their Discoveries, San Francisco: Freeman, 1980.

Serres, M. (editor), A History of Scientific Thought, Oxford: Blackwell, 1995.

Solomon, Joan, The Structure of Matter: The Growth of Man's Ideas on the Nature of Matter, New York: Wiley/Halsted, 1974.

Stroke, H. Henry (editor), The Physical Review: The First Hundred Years. A Selection of Seminal Papers and Commentaries, Woodbury, NY: AIP Press, 1995.

Sui, H., Han, B. G., Lee, J. K., Walian, P., and Jap, B. K. (2001). Structural basis of water-specific transport through the AQP1 water channel, Nature 414, 872-8

Suplee, Curt, Physics in the 20th Century, New York: Abrams, 1999.

Tajkhorshid, E., Nollert, P., Jensen, M. O., Miercke, L. J., O'Connell, J., Stroud, R. M., and Schulten, K. (2002). Control of the selectivity of the aquaporin water channel family by global orientational tuning, Science 296, 525-530

Tomonaga, Sin-itiro, The Story of Spin, translated by T. Oka., Chicago: University of Chicago Press

Torrance, John (editor), The Concept of Nature, New York: Oxford University Press, 1992.

Toulmin, Stephen, and Goodfield, June, The Architecture of Matter, New York: Harper, 1977.

Toulmin, Stephen, and Goodfield, June, The Fabric of the Heavens, New York: Harper, 1965.

Truesdell, C., Essays in the History of Mechanics, New York: Springer-Verlag, 1968.

Van Helden, Albert, Measuring the Universe: Cosmic Dimensions from Aristarchus to Halley, Chicago: University of Chicago Press.

Weart, Spencer R., and Phillips, Melba (editors), History of Physics: Readings from Physics Today, Number Two, New York: American Institute of Physics, 1985.

Weaver, Jefferson Hane (editor), The World of Physics: A small Library of the Literature of Physics from Antiquity to the Present, three volumes, New York: Simon and Schuster, 1987.

Westfall, Richard S., The Life of Isaac Newton, New York: Cambridge University Press, 1993.

Wheaton, Bruce R., The Tiger and the Shark: Empirical Roots of Wave-Particle Dualism, New York: Cambridge University Press, 1983.

Whitehead, Alfred North, Science and the Modern World, New York: Free Press, 1967.

Whittaker, E. T., A History of the Theories of Aether and Electricity, Volume I, The Classical Theories; Vol. II, The Modern Theories, New York: Tomash Publishers and American Institute of Physics, 1987.

Yoder, Joella G., Unrolling Time: Christiaan Huygens and the Mathematization of Nature, New York: Cambridge University Press, 1988.

Zajonc, Arthur, Catching the Light: The Entwined History of Light and Mind, New York: Oxford University Press, 1993.

Zhu, F., Tajkhorshid, E. and Schulten, K. (2003). Theory and simulation of water permeation in Aquaporin-1. Biophysical Journal, 86, 50-57

10 INDEX

Abell 2667
    galaxy cluster, 107
actin, 119
American Association for the Advancement of Science, 153
American Museum of Natural History, 153
aquaporins
    discussion, 72
    found in diverse organisms, 73, 110
archaeological dig
    quest, 23
artificial intelligence, 125
Asimov, Dr Isaac
    quote on human brain, 82
astrocyte cells
    multi-functional nature, 83
black holes, 100
bladder
    Darwinian principles, 111
blood
    oxygenating, 114
blood vessels
    illustration of fractal branching, 65
    length, 114
Blue-ray discs
    DNA comparison, 126
bone
    example of Dynamic Evolution, 43
brain
    and computers, 125
    blood-brain barrier, 84
    cell migration in embryo, 82
    homeostasis, 84
    lemniscus, 87
    neurons, 54
capillaries
    and nerve fiber development, 87
    average length in human body, 88
    illustration of branching of blood vessels, 88
    patterns of tubes, 111
cardiovascular system, 111
Carl Sagan
    brain quote, 74

cell
    illustration, 104
    traffic control for, 43
chaetopterid worm, 120
Chicago conference, 153
chicken and egg question
    DNA, 90
Cinipid Wasp, 120
cochlea, 119
computers
    language of, 126
configurative connections, 110
    fractal patterns of the plexus, 93
consequential entities
    bulleted list of, 50
    categories, 42
    definition, 29
    dependencies, 29
    fine-tuned entities, 29
    outline, 27
    relationship types, 47
    relationships, 45
    spleen, 48
    synergism, 30
cytokinesis, 118
Darwin
    On the Origin of Species, 13
    Origin of Species, 16
Darwinian Evolution
    anthropomorphism, 74
    is there a "better fit"?, 13
    portraying skull sizes, 143
Darwinian Model
    passage of great amounts of time, 16
    passivity, quiescence, 20
    strong nuclear force, 46
digestive system, 120
disparate functionality, 111
DNA
    and computers, 125
    discussion, 95
    illustration, 103
    links to other entities, 78
    multi-functional, 80
    storage capacity, 125
domino effect, 30

Dynamic Evolution
    approach and methodology, 20
    dynamic vs passive, 19
    multiscalar approach, 21
    outline, 71
    scope, 19
earth
    magnetic shield, 100
earth-moon system, 100
    illustrated, 105
eccentricity diagram, 49
Einstein
    quote, "I want to know his thoughts", 22
electromagnetism, 100
    patterns, 110
elephant
    unborn baby illustration, 97
embryo
    example of Dynamic principles, 93
    illustrated, 95
    neurovascular development, 86
Encyclopaedia Britannica
    spleen reference, 48
endoplasmic reticulum
    multi-functional properties, 80
energy output
    staged, 99
entities. *See* Consequential Entities
epimysium, 118
ethylene
    molecule illustration, 102
Everett Olson
    dim view of fossil record, 154
eye
    development in womb, 89
    example of Dynamic Evolution, 44
    involuntary blink, 44
    tear ducts, 44
fascicle, 118
fetus
    pattern for, 98
fibonacci numbers, 53
fine-tuned entities, 29
fly (common), 120
fossil record
    quotations, 145
galactic wall

patterns of the Plexus, 100
galaxy cluster
    illustrated, 107
glial cells
    multi-functional, 81
gluons, 99
    cloud, 100
    patterns, 110
glutamate, 84
golden ratio, 53
Gould, Stephen Jay
    microevolution quote, 154
grand unified theory of everything, 19
gravitons
    patterns, 110
gravity
    patterns, 110
heart
    illustration in rib cage, 105
    patterns of tubes, 111
Isaac Newton
    seashore quote, 15
lemniscus, 87
liver, the
    multi-functional characteristics, 85
lungs, 120
lymphatic vessels, 86
Macroevolution, 153
mammary glands, 120
Matryoshka Doll, 116
mesenchyme, 86
mesoderm, 86
metabolites, 84
Milky Way galaxy
    illustrated, 106
mosaic
    picture paints a thousand words, 17
moth, noctuidae
    example of Dynamic principles, 80
Murray, Prof. Sir Robin
    quote on human brain, 82
muscle fibers, 115
myofibril, 118
myofilaments, 118
myosin, 118, 119
natural selection

comparison to embryonic development, 93
nervous system
    development, 84
neurotransmitters, 84
neurovascular development
    multi-purpose, 86
Newton, Sir Isaac
    quote, boy playing on seashore, 15
Niles Eldridge, 153
noctuidae moth, 80
octopus
    human-like eye, 154
pancreas, 120
patterns
    embryonic, 98
    proteins and cells, 99
    seen in babies and children, 95
    sub-atomic particles, 99
Penrose, Sir Roger
    quote on human brain, 82
periodic table, 53
photons
    patterns, 110
photosynthesis
    dependent on Entity water, 34
plexus
    4D model, 42
    connection types, protein example, 47
    fractal reticulation, 65
    illustrated, 41
    syzygy, 53
pollen tubes, 111
polychaetes, 120
proteins
    discussion, 47
    folding, 99
pulmonary artery, 114
quantum
    particle collisions illustration, 101
quantum mechanics
    and computers, 125
    and Dynamic Evolution, 28
radial glial cell, 81
rectum
    Darwinian principles, 111
reproductive system, 120
Russian Dolls, 115

manufacture, 119
 salivary gland, 120
 sea angel, 120
 sieve tubes, 111
 skeleton
  development in womb, 89
 solar system, 100
  illustrated, 106
 space shuttle
  example used, 75
 spinal cord, 84, 89
 stomach
  Darwinian principles, 112
 strong nuclear force
  fine-tuning, 46
  patterned with gravity, 110
 superclusters
  illustrated, 108
 survival of the fittest, 13
 Thurston lava tube, 113
 toroidal geometry, 121
 tracheid cells, 111
 tubes
  patterns in nature, 110
  used in space shuttle, 110
 water
  dependencies - illustration, 37
  dependents - illustration, 33
  list of properties, 33
  molecular interaction illustration, 46
  molecules illustration, 102
  polarity of the molecule, 43
  single water molecule, 31
  surface tension, 35
 zygote
  DNA pattern, 98

www.ingramcontent.com/pod-product-compliance
Lightning Source LLC
Chambersburg PA
CBHW060845170526